SENSORS
in highway and civil engineering

SENSORS
in highway and civil engineering

Proceedings of the conference organized by the Institution of Civil Engineers, London, 5 February 1981

THOMAS TELFORD LTD, LONDON, 1981

Published by Thomas Telford Ltd for the Institution of Civil Engineers, Telford House, PO Box 101, 26-34 Old Street, London EC1P 1JH

ISBN: 0 7277 0109 6

Printed by Tetford Press and bound by Norton Bridge Bookbinding Ltd, Stotfold, Hitchin, Herts.

CONTENTS

1

Laser control of construction equipment

A. TWYNING, Technical Support Engineer, Spectra-Physics Ltd

The development of the continuous gas laser, particularly the Helium Neon has made possible new methods of grade control for construction equipment. As all the sensing equipment we will discuss depends upon this laserlight let me briefly talk about its properties. The laser beam is:

(A) Mono-chromatic, one colour, in this case red, with a wave length of 632.8 nanometers.

(B) Coherent. Light is only one single frequency in phase amplitude and direction throughout its path.

(C) Low divergence. The beam spread is very low. 1.0 milliradiums.

(D) Intensity loss is also low compared to conventional light. Atmospheric absorbsion being only 5% and total loss over 300 metres being about 65%. (1)

The last two properties are the main ones of direct interest to us. A raw spot when generated by the plasma tube, the heart of the laser transmitter, is quite small, .159 centimetres, which is difficult to see and sense. With simple optics we can enlarge this beam up to 1.5 centimetres diameter and arrange collimation to keep the beam reasonably constant over 300 metres. (2)

The early lasers used in pipe laying provided a reference beam of light which could be set at a desired gradient and the beam was observed by some form of target. This ranged from a purpose made translucent target, which fitted in or over the pipe to the operator's wellington or shovel handle. By rotating the beam and generating a plane of light a more sophisticated sensor is required, but gives much more versatility. (3)

Functional Description

The Model 975 Laser Eye Detector is the world's most advanced compact detector. Its separate sensor and display and mounting accessories have been designed for maximum versatility in mounting and beam detection. Two mounting clamps, a battery charger and a carrying case are included with the Model 975.

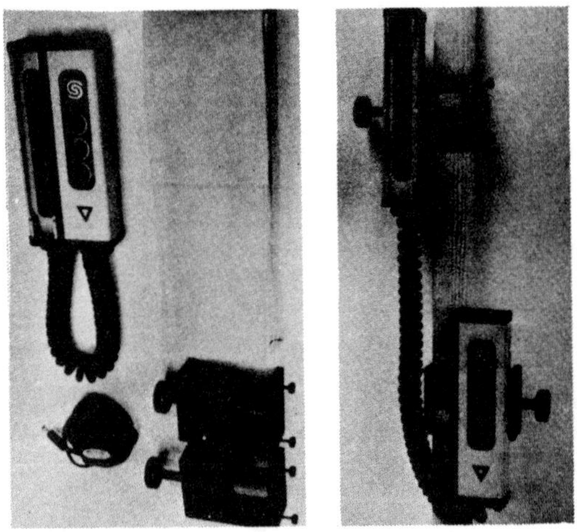

Applications

Concrete flatwork, general layout, sitework, highway slope control, footings, elevation control.

Operating Indicators/Controls:

Marking Notches - Four edges of the sensor unit are notched to allow easy marking of forms and walls.

Three Light Beam Position Readout - Indicates when sensor is high, low, or right on grade.

Sensitivity Switch - Controls sensitivity of the on-grade indication (two positions).

On/Off - Supplies electric power to unit.

Fig. 1. Model 975 laser eye

A simple detector such as the Spectra-Physics Model 975 Laser Eye (TM), (Fig. 1), gives the operator an indication of high, low and on-grade, over a search range of 152 mm. If the rotating plane of light is horizontal, then levels can be quickly and easily transferred, or by coupling the read out of the detector to a calibrated staff, general surveying work can be carried out. These operations can be further speeded up by including the sensor with the staff in one unit. The Spectra-Physics 970 Laser Rod (TM) (Fig 2), automatically moves the sensor over 9 metres, or 2.2 metres, to locate the reference beam. These operations allow one man very quickly to transfer levels and to perform certain surveying tasks, but as this is a subject in itself, it will not be expanded here. (4)

The success of the compact Spectra-Physics 944 Laser Level (TM), (Fig.3), which gave the horizontal plane of light was further developed so that a plane of light could be inclined to a percentage grade setting. It still retained the features of self-levelling within plus or minus 8%, and only allowed the beam to be switched on when the desired level reading was obtained, preventing misreadings. If a receiver is connected to the blade of a bulldozer, scraper, grader, or excavator, then exactly the same information of high, low, on-grade with respect to our reference plane can now be given direct to the operator. (5)

The Spectra-Physics Laserplane System provides a variety of equipment for laser machine control, and I want to mention here only the simplest. This is the indicate system, which has a large display, showing the green on-grade condition, with yellow arrows indicating high, or low. Extra information is provided by the receiver, the outer segments display a steady off grade indication and as the on-grade position is approached, the lights pulse helping to eliminate over-shoot. This system as the name implies is just an indicate system, and gives the operator information as to his position to the required grade previously set into the transmitter. This is the simplest and so less expensive system available, especially when considering installation costs to the machine. The main consideration is where to fix the receiver, the general rule is above the cutting/operating edge. The receiver slides on to a rigid mast 4.45 cm diameter, and typically 1.22 metres long, and can be bolted direct to the machine if convenient, or to a welded bracket.

Although it looks like a common surveyor's grade rod, the Model 970 Direct Reading Laser Rod contains a laser sensor. This sensor tracks up and down the Laser Rod until it automatically locates the laser beam. The increased speed and accuracy provided by the Model 970 has revolutionized the field of construction jobsite measurements.

The Laser Rod is actually two rods in one. You can set it to read like an ordinary Philadelphia rod, or you can set it to read "direct" elevations above sea level. The ability to read elevations directly eliminates field computations - saving time and reducing errors. You can use the Laser Rod with a Spectra-Physics self-levelling Laser-level in applications such as:

*Contour surveying for site work.
*Setting grade stakes for building sites, roads, airports.
*Levelling footing batter boards.
*Setting stringlines for graders and pavers.
*Marking piling cutoffs.
*Checking sub-base grades.

The Laser Rod gives you measurement numbers at the point where you need them - not hundreds of feet away at an instrument. This will save you labour cost of more than 50% over previous methods, and provide you with increased productivity, accuracy and profits on your construction jobsites.

Fig. 2. Model 970 direct reading laser rod

The Spectra-Physics Model 945 Laser-level GS is the world's most advanced laser tool, designed to increase productivity of general and heavy contractors. It is an electronically self-levelling rotating laser that provides a plane of laser light which can be aligned and stabilized at slopes of up to 10% from true horizontal. This capability is revolutionizing many aspects of construction alignment and control.

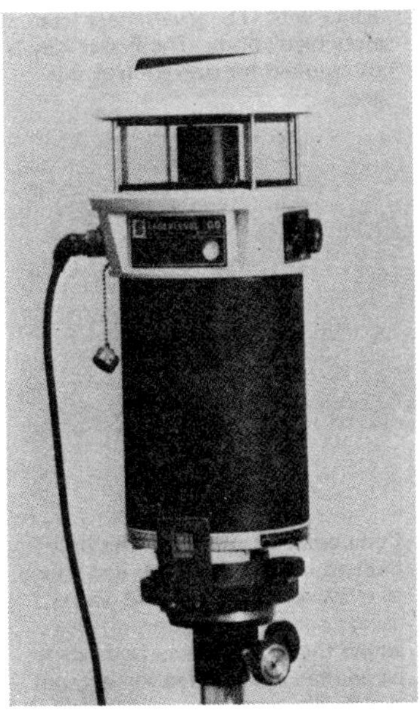

Applications

This laser, in conjunction with a laser detector (The Spectra-Physics Model 970 Laser Rod) can be used as a one-man construction crew for tasks such as:

* Setting and marking grade stakes.
* Setting alignment guides for slip-forming machines.
* Installing string lines for auto-graders and paving machines.
* Checking auto-grader performance.
* Setting concrete screeds.
* Marking piles for cut-off.
* Setting and verifying highway cross-slopes.

This laser can be mounted on its side to provide a self-plumbed vertical plane for:

* Plumbing wall and bridge forms.
* Layout of stake lines.
* Aligning vertical support members.

Fig. 3. Model 945 laser-level GS

The display is also bolted on, and can be positioned for best convenience to the operator. (6)

Power is 12 v D.C. and can be taken from the machine's own battery. It can be seen that this is a very easily installed system and so can be transferred from machine to machine very quickly. (7)

To operate the machine the blade is lowered on to the bench mark, or level required, the receiver which slides up and down the rigid mast is positioned so that the on-grade light is displayed. The operator then continues as normal, until he receives a steady indication from the display. He then knows he is within 8 cm to desired level. A flashing indication starts when the blade is 4 cm above desired level, and finally a green on-grade light comes on when the blade is the same height that was initially set. The advantages of this system is that the operator knows instantly if he goes below the desired grade, thus preventing time consuming infill at a later date. The operator now has a continual monitor which can enable him to grade an area flatter and quicker, without constant reference to outside grade checkers or profiles. If required the system can even be operated at night or poor light conditions. (8)

This basic system can be up-graded with the addition of an electric mast to replace the rigid mast making initial set up quicker, in that the operator does not have to leave his cab to set the receiver height. The system can also be up-dated to provide a signal which will control an electric valve which can be interfaced with a machine's hydraulics to provide automatic grade control. (9)

Automatic grade control gives a further sophistication in that due to the speed of the laser rotation of 300 revolutions per minute, the machine's hydraulics will be corrected five times a second, much faster than any operator can achieve. For very fast moving machines this can be increased to 600 revolutions per minute to give 10 corrections per second. (10)

Automatic grade control for construction equipment is now well established in the U.S.A. Europe, Middle East and Australia, but at this time in the U.K. it is only being used on land drainage machines. Perhaps the British conservatism, lack of large land levelling projects, good operators and costs are factors why we in the U.K. are slow to accept the concept. This is why the indicate

Three Light Display

The Three Light Display is an electronic device which graphically indicates, by means of lighted indicator segments, the position of the laser receiver relative to the plane of laser light. A centre green rectangle lights to indicate "On Grade". Yellow Up and Down arrow segments are lighted to indicate that the receiver is below or above grade. The unit can also drive solenoid valves in an On/Off (PT) system.

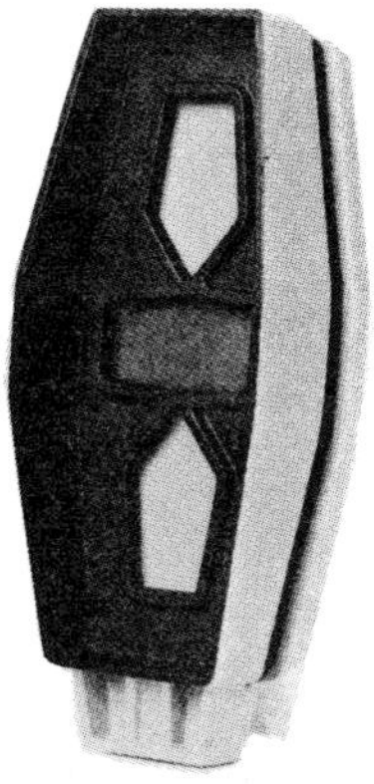

Fig. 4. Three light display

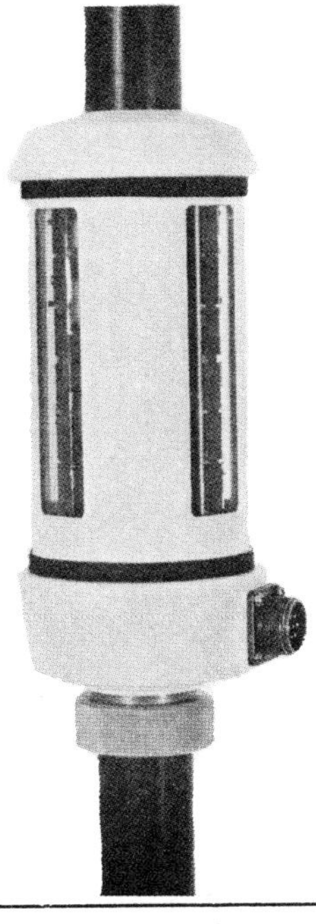

Five Channel Receiver

The Five Channel Receiver is a photo electric sensor that detects the plane of laser light as it strikes the photocell stack. The receiver then signals the control station the position of the plane of light relative to the "On Grade" or centre position of the photocell stack. The receiver has four stacks of photocells 90° apart which provide for full 360° sensing range. Vertical sensing range is 7 in (17.7 cm).

Fig. 5. Five channel receiver

Fig. 6. Bulldozer: three light display system with electric mast

Fig. 7. Paver: three light display system working automatically to control level of paver. A receiver mounted horizontally to the steering leg senses a vertical beam to control the direction automatically

Fig. 8. Grader: three light display system fitted to grader

Fig. 9. Excavator: three light display system fitted to excavator. Quick release shock mounts are shown. True reading of depth is obtained when boom is vertical.

system is, I think the answer. (11)

It can be used for small jobs such as warehouses, car parks, footings and sports fields. It utilises the good operator and is less expensive than the automatic system. The same indicate display system, versitility is such that it can be fitted to dozer, scraper, grader, excavator and paver. (12)

What accuracy can we expect from the indicate system? Obviously it depends on the skill of the operator, but the laser provides a reference within 2mm over 30 metres, and the receiver has an on-grade dead band of 6mm. (13)

Figures 4 to 9 are pictures from job sites in the U.K. (14)

2

Sensors for road pavers

T. R. SHELLEY, MA, PhD, MIMM, Research Officer, Babcock Blaw Knox

SYNOPSIS. The floating screed action of pavers is explained. It is shown that there is a considerable lag between a change in the tow point position and the depth of material being laid. This affects sensor technology. The current sensors are described as are the different methods of using them. The problems of using more sophisticated systems are summarised and a plea is made for the development of suitable sensors. Some of the company's experiences with regard to laser systems are reported.

1. The paving screed is pulled along from a point several metres in front of it. It is pulled through and supported by the material being laid. The material is partially compacted by the screed and this results in it being inclined to the horizontal by an angle known as the "Angle of attack", Fig.1. For proper operation, the screed should be supported within the material, and not by the road base directly through the material: in other words, it should not exceed the bearing pressure of the material. As a result, at significant depths, the angle of attack is, to a first approximation, independent of the depth being laid. In fact, it increases slightly with depth but this effect is ignored in the following calculation.

2. <u>Calculation of response to a change in tow point</u>. If the side arm length is l, and the tow point is raised by a height h, in response to a deviation between the reference level and the thickness of the mat being laid, the angle of attack of the screed is increased by $\tan^{-1}(h/l)$ and it therefore, begins to plane upwards at this angle. However, after the paver has moved torwards a distance x, the screed is at height y, above its original position. However, the angle at which the screed is planing up is now only:

Sensors in highway and civil engineering. ICE, London, 1981, 11-18

$$\tan^{-1}\left(\frac{n - y}{l}\right)$$

As the screed moves forward δx, it rises upwards a further distance, δy where:

$$\delta y = \left(\frac{h - y}{l}\right)\delta x$$

Hence as $\delta x \rightarrow 0$

$$dy = \left(\frac{h - y}{l}\right) dx$$

$$\int \frac{dy}{(h - y)} = \int \frac{dx}{l}$$

$$\ln (h - y) = k - \frac{x}{l}$$

But when $y = 0$, $x = 0$

Hence: $\ln \left(\frac{h - y}{h}\right) = -x/l$

And so $y = h\left(1 - e^{-x/l}\right)$

3. This formula is not completely obeyed in practice as the new stable angle of attack is greater than the previous angle of attack, if the depth of material under the screed is increased and less if the depth is decreased. This cuts off the observed response curve at approximately a side-arm's length.

4. The inherent sluggishness to respond is desirable in that perturbations in the position of the tractor due to bumps in the road base are not transmitted to the mat being laid. However, there are consequently inherent problems when the screed is being made to follow a reference level. The solution is to put the sensor monitoring the reference level in advance of the screed. However, if the sensor is at say, the tow point, the tow point height will be accurately controlled but there will be no input from where the screed is and what thickness of mat is actually being laid. If, on the other hand, the sensor is near the screed, the time for the screed to respond to a change in tow point will not be taken into account. Care must be taken that when the machine stops with a deviation between sensor and reference level, the tow point ram does not continue to move up or down without being able to produce any change in depth of mat. One common solution is to position the

sensor between the tow point and the screed arm. It is either mounted on the side arm or on a beam extending forwards from the screed end.

5. Blaw Knox Sensors. The present sensors that Blaw Knox offer for automatic control are simple and especially rugged devices intended to be compatible with the environment they have to survive in. More sophisticated techniques using precision optics and/or electronic systems are being studied, but it is difficult to find any device which will survive severe vibration, rain and the general abuse typical of a construction site. Vertical deviations are picked up by a swinging arm device and microswitches which in turn actuate the solenoid valve to the tow point ram in the appropriate direction. This device is designed to reference either to (i) a kerb or section of mat already laid by being equipped with a shoe, (Fig.2.); (ii) a reference wire: (iii) an averaging beam (Fig.3.).

6. The other Blaw Knox made device is the crossfall box (Fig.4.). This contains a heavy oil-damped pendulum which contacts one of two microswitches when the crossfall between the sidearms is greater or less than a pre-set amount. A common means of using this is to use a level sensor to control one tow point and the crossfall box to control the tow point on the other side away from the reference level (Fig.5.).

7. Long Grad-line and other systems. Blaw Knox also offers Long grad-line equipment with their machines. Other equipment can also be fitted which has similar facilities. These incorporate a "dither". Small corrections in tow point position can be made by pulsing the valve spool. Larger corrections (7mm or more on the level sensor) will begin with a continuous signal and change to a pulsed signal as the correction nears completion. Honeywell produce a time proportioning system which similarly produces a variable rate of response according to the magnitude of the deviation in order to prevant the overcorrection which is possible with bang-bang systems.

8. It has been shown possible to use these systems to link two pavers in echelon. Such a system is known as "twinning". This was tried, for example with two PF980 pavers laying the A9 autoroute near Perpignan, France in 1975. The leading paver in the left hand lane had a Honeywell level sensor working on a string line to control the left hand tow point. The trailing paver had a Blaw Knox level sensor

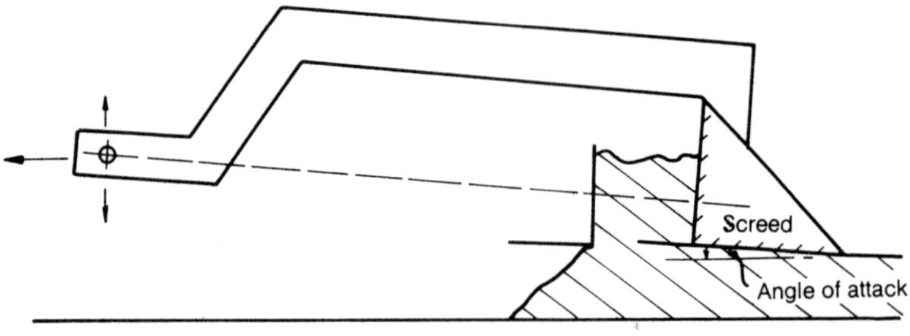

Fig. 1. Floating screed action

referencing to the right hand string line and controlling its right hand tow point. The left hand tow point was controlled by a second Blaw Knox level sensor referencing to the right hand edge of the mat laid by the leading paver. Both pavers had Honeywell slope controls, the signals from which were fed to a control box which averaged them and used this signal to control the right hand ram of the leading paver. The leading machine was fitted with Honeywell proportional control valves while the trailing paver was not. The arrangement was quite successful and neither machine was stopped if the other ran short of material.

9. Possible sophisticated aids. There are a number of other possible systems which could be used. Any system has to be able to withstand considerable machine vibration, any kind of weather, and being thrown in the back of a van or hopper at the end of the day in a puddle of water. It must also not cost too much. It would be desirable to do away with shoes on level sensors which can be knocked off by drain covers etc., and string line pickups which can come off the wire. To do this requires either non-contact sensors accurately monitoring the distance of a moving surface less than a metre away or the position of a transmitted reference beam. If there are any suitable non-contact surface sensors capable of fulfilling the necessary requirements of ruggedness at low cost, the author would like to hear about them. As regards the beam sensing sensors, they have problems all of their own.

10. Laser beam systems. There is a misconception that laser beams are necessarily perfectly straight. Laser light is subject to refraction by volumes of air at different temperatures in the same way as any other kind of light. In addition to this, laser light beams do diverge slightly.

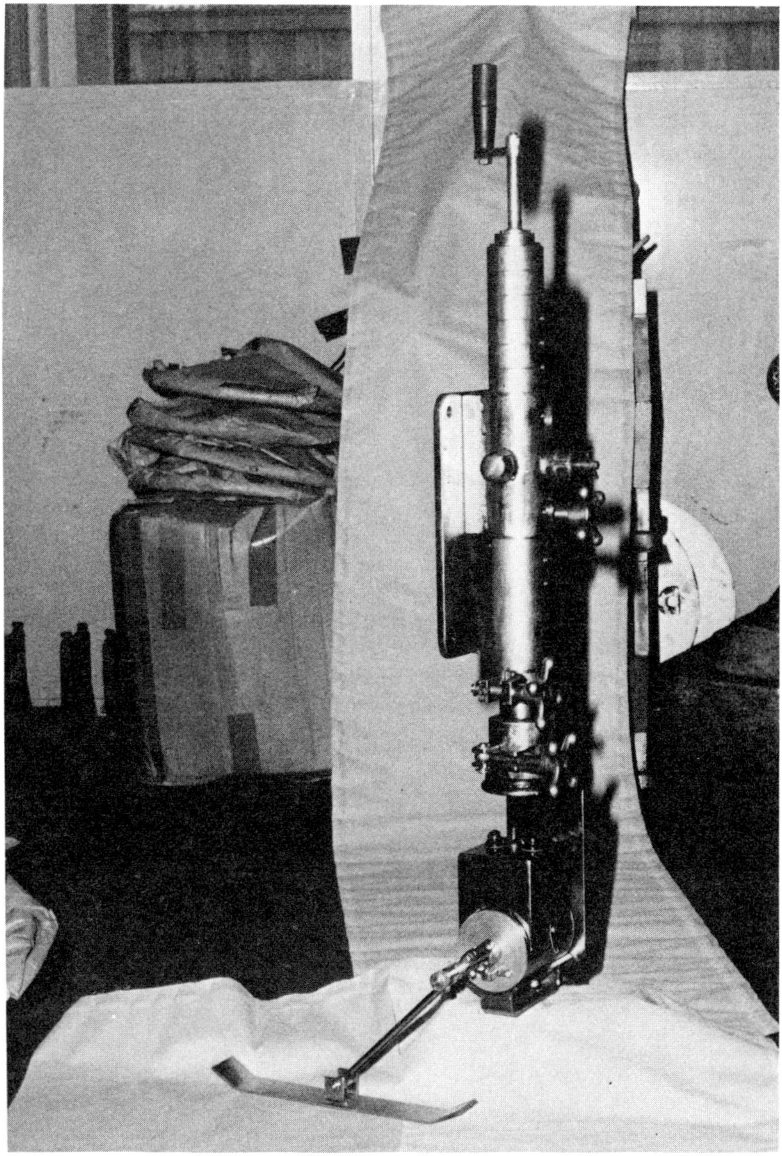

Fig. 2. Blaw Knox level sensor

Fig. 3. Averaging beam on BK95 paver

Fig. 4. Crossfall box on PF90 (The box itself is immediately behind the engine cowling).

11. If using laser plane device, where the beam is rotated like a lighthouse beam, the limiting distance is 500m under ideal conditions. The vertical accuracy is said to be about 0.01% i.e. 1mm in 10m (ref.1.).

12. A trial was held in February 1980 where a laser plane was used to maintain levels for grading and paving with lean concrete on a large site at Ramsgate Harbour. The system was certainly little trouble to use but fluctuations in the received level were unacceptable a few hundred metres from the level. If asphalt had been being laid, the convection currents produced by hot asphalt would have greatly increased this fluctuation. Trials in France have confirmed that it is not practicable to use lasers to maintain reference levels for pavers laying asphalt because of convection effects. These effects could be reduced by using a longer wavelength than that produced by the He-Ne lasers popular in the construction industry. A possibility is the 10.6 µm produced by CO_2 lasers, but they suffer from the problem of being too powerful, producing 2 watts or more of light when the maximum permissible

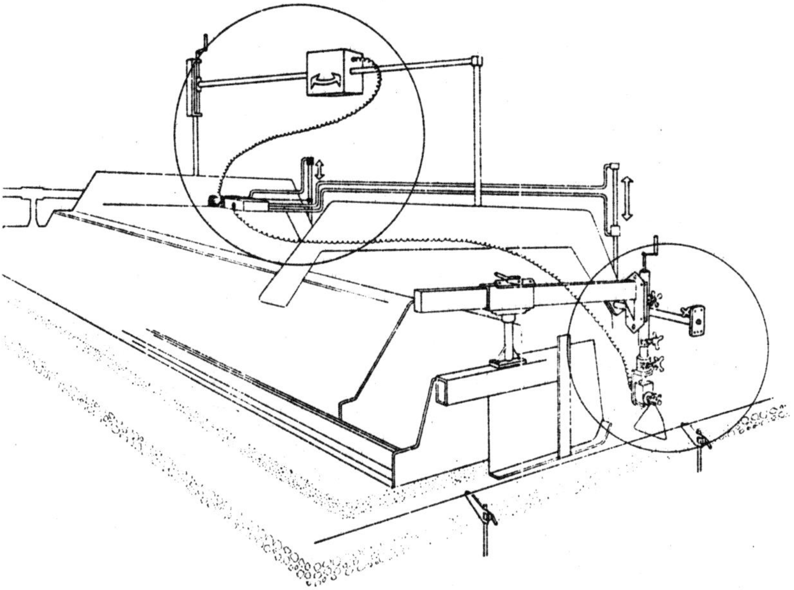

Fig. 5. Stringline controlling right hand side and crossfall box controlling left hand side.

from the safety point of view would be of the order of 10 mW. A possible solution would be to use a modulated infra-red L.E.D. with conventional optics to produce the narrow beam. The modulation would be necessary in order to distinguish between the level source and stray infra-red light. Such a system has yet to be developed.

13. A further possibility would be to use millimetric microwaves instead. However, again, a suitable system has yet to be developed. The generator alone would cost about £10,000.

CONCLUSION

14. Servo valves and micro-electronics make it relatively simple to produce a machine which could lay road material in any manner desired. However, this is futile unless the machine is fully informed as to where it should be going and how well it is doing. The time may come when an operator starts his machine, and the paver computer interrogates him, (with a choice of voice, language and personality), as to the width and thickness of mat desired, course of road and banking of curves. At the appropriate word of command the machine could obey, pausing only for material, large obsticals or fuel until told to stop. This technology is possible but it will be successfully applied only when suitable sensors become available cheaply.

REFERENCES

1. CLARKE R.J. Lasers: their uses, benefits and limitations in the construction industry. Highways and Public Works, 1980, 48 April, 6 - 8.

3

Dynamic assessment and control of piling operations

J. N. MURE, BSc, MICE, and M. L. KIGHTLEY, BSc, Balken Piling Ltd

INTRODUCTION

The control and assessment of driven piling operations has traditionally been attained by the following methods

A. Site investigation survey prior to and occasionally during piling operations on site in order to assess soil characteristics.

B. Driving formulae to obtain some practical means of assessing pile capacity.

C. Full scale load testing, sometimes known as static load testing, either to failure or some lesser load, in order to assess pile capacity and/or to indicate settlement in service.

There are cases where all three methods are used and also cases where only one or two of the methods are used.

More recently the wave equation has been practically applied as a means of assessment and control, and techniques have been developed whereby dynamic measurements of stress wave propagation can be simply carried out on the job site and used as a control for piling operations.

This paper describes the Case Method for dynamic measurements of stress wave propagation used as a system for controlling and assessing pile driving operations in respect of three work sites in the United Kingdom where Balken precast concrete piles were installed.

Sensors in highway and civil engineering. ICE, London, 1981, 19–34

The Case Method

The Case Method is so named after Case Western Reserve University Ohio where a twelve year research project was carried out under the direction of Professor G. Goble and sponsored by the Ohio Department of Transportation.

The research project was aimed at finding a reliable and accurate means of assessing pile bearing capacity by carrying out dynamic measurements on driven piles at installation. During the course of research, it was found that other interesting properties of piles during their installation could be observed.

Originally the pile was assumed to be a rigid body of mass m, pile top force and acceleration were measured as functions of time and soil resistance R could be computed as

$$R = F(t) - m.a(t) \qquad (1)$$

The solution was not practically applicable because elasticity of the pile has to be considered and the following equation was derived

$$R = \frac{F(t_1) + F(t_2)}{2} + \frac{MC}{2L}\left(V(t_1) - V(t_2)\right) \qquad (2)$$

t_1 is the time of impact or max. pile top velocity

t_2 is at a time interval of $t_1 + \frac{2L}{C}$

C is the wave speed for the pile material

L is the pile length from pile toe to measurement gauge attachment point

V is velocity of the pile top obtained by integrating acceleration.

Further

$$R = R_s + R_d \qquad (3)$$

R_s = static resistance (pile capacity)

R_d = dynamic resistance due to penetration of the pile into the soil.

The dynamic resistance or damping resistance is proportional to the velocity of the pile toe and is equated by means of a soil dependent factor J_c, the Case Soil constant which is a dimensionless damping constant

$$R_d = J_c \frac{EA}{C} V\,toe \qquad (4)$$

The pile toe velocity is calculated

$$V\,toe = 2.V_{top} - \frac{L}{m.c.} R \qquad (5)$$

By subtraction of dynamic from total resistance the static resistance or pile capacity can be determined.

The damping constant J_c is dependent on the characteristics of the soil at the pile toe and increases in value with decreasing soil particle size. Values for J_c have been obtained by direct comparison of dynamic and static failure tests of a large number of piles driven in various soil types.

When assessing pile capacity by the Case Method it is necessary to be able to substitute a value for J_c. This can be obtained by two means: site investigation taken to a depth below final pile toe level or statically testing pile to failure in order to determine J_c for a particular site.

Equipment

Strain transducers and accelerometers are bolted by stud anchors to concrete pile surfaces close to the pile head. These gauges transmit signals of strain and acceleration during pile driving for each hammer blow. The signals are accepted by the Pile Driving Analyser which converts the strain signal to a force and by integration the acceleration to a velocity.

The unit is used in conjunction with an oscilloscope which displays the force and velocity traces on a time base for each hammer blow and an instrumentation tape recorder which retains the record of force, acceleration and spoken comments as a permanent record and for further use in analysis off the site.

The pile driving analyser determines the following properties

pile capacity
measurement of piling equipment performance
checks pile integrity
energy transfer to the pile
compressive and tensile forces
maximum pile velocity and acceleration.

Assessment of pile integrity

As an integral part of dynamic tests carried out on Balken piles using the pile driving analyser an integrity check is performed.

For every hammer blow a curve of force and velocity is displayed on the oscilloscope. The force and velocity are related by the pile impedance ($\frac{EA}{C}$) and this relationship is the basis for the checking of pile integrity.
The relationship of force and velocity breaks down when a wave is reflected to the pile head before the primary stress wave is reflected from the pile toe to the pile head. This effect shows up at the oscilloscope as a change in the proportionality of the force and velocity curves.

Such effects can be caused by :

increased pile cross section
reduction in pile cross section
transverse crack in pile

The form of the change in proportionality identifies the different causes.

Refer to Figs 1 and 2 and a change in proportionality is noted. This has been caused by a "slack" in the second pile joint (the joint nearest the pile head) and this has been measured as 0.15 mm in width. Note that a transverse crack in the pile would appear as a similar effect.

Site procedure

Normally the analyser is used in two ways

A. a test can be made from start to finish of

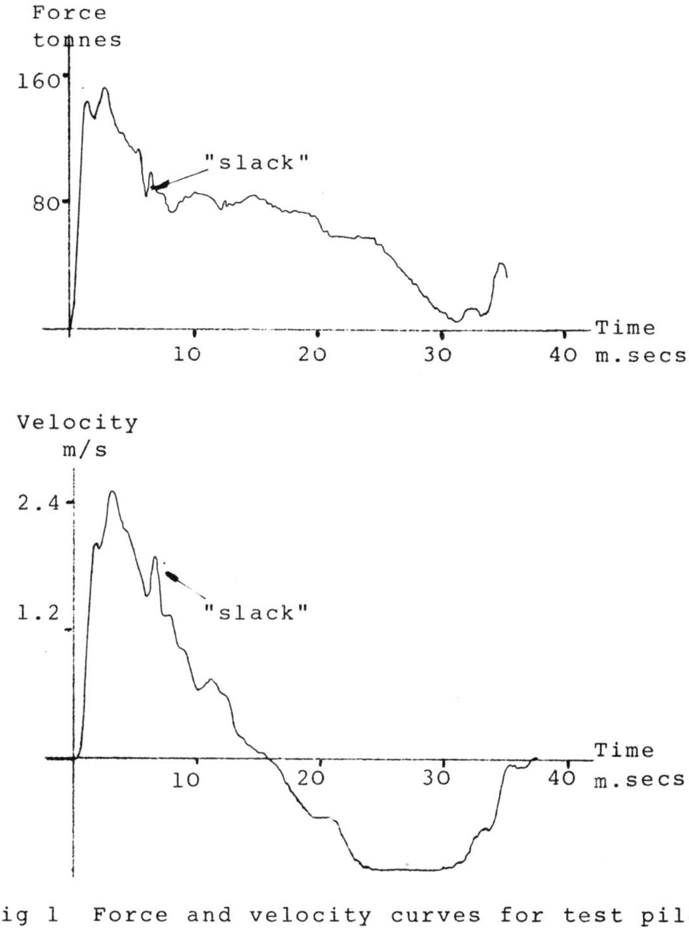

Fig 1 Force and velocity curves for test pile at end of installation South East Lancs. Note "slack" at pile joint.

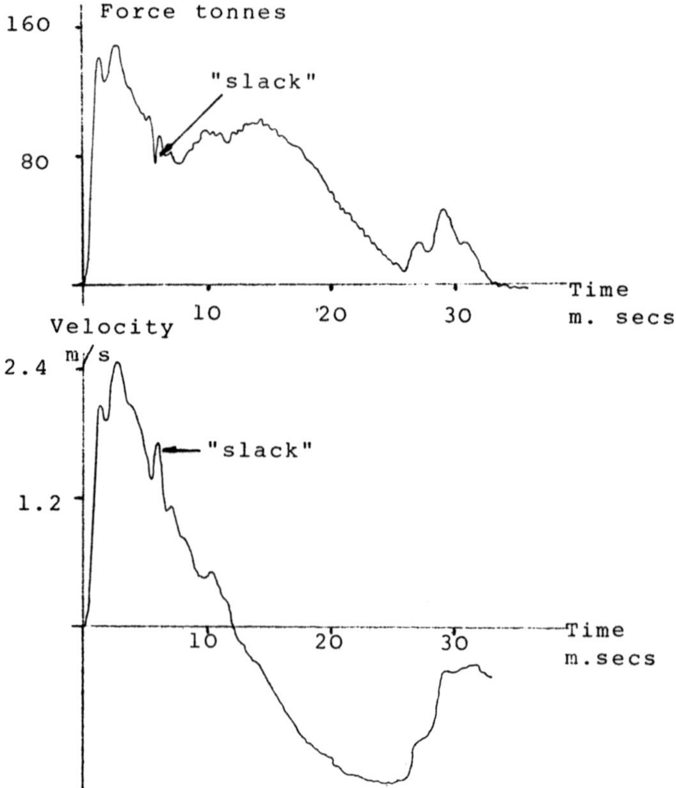

Fig 2. Force and velocity curves for test pile
1 day after installation

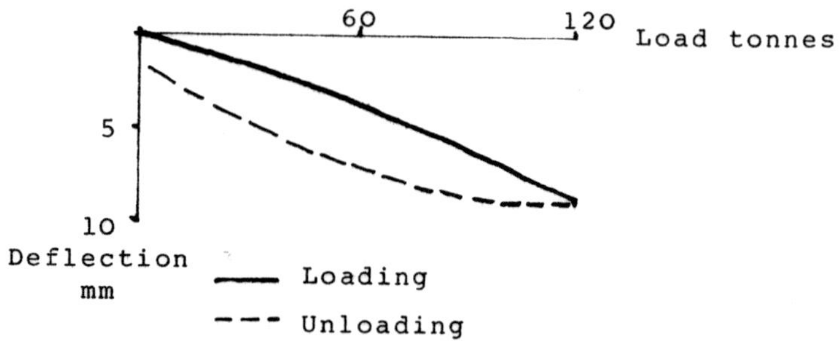

Fig 3. Load.V.Deflection for test pile.

pile installation giving a complete record. This record is of interest in assessing the performance of the driving equipment and in observing the maximum forces acting on the pile at all stages of driving. Also, pile capacity at any stage of driving can be determined and any required adjustments to installation procedure, depth of installation or driving equipment can be made if required. Our experience has shown that a complete record is useful at job beginning in order to make such assessments and possibly set out driving criteria.

B. a test can be carried out at some time later to pile installation - known as test at re-strike. To carry out such testing a piling rig is required and piles to be restruck must be accessible to the rig. Restrike testing is interesting in observing pile integrity and also pile capacity with respect to any take up or relaxation effects on capacity with time. Testing in this way means that a relatively large number of piles can be tested in one day.

Case studies

It should be noted that all the piles tested in the following studies were standard Balken Type 1 precast concrete piles of 275mm square section.

1. South East Lancashire Some one hundred piles were driven to support a road bridge crossing a new by-pass. The site investigation indicated very soft clay to 4 metres depth, sandy silts and silty clays to 18 m depth underlain by firm becoming stiff silty clays. Piles were installed to approximately 30 metres length to attain a minimum ultimate load of 120 tonnes to give minimum factor of safety of 2 over working load.

The specification called for two piles to be statically load tested. It was suggested as an alternative that the first test could also be dynamically tested and the second load test dispensed with in order to test a greater number of working piles dynamically. This was accepted in principle, pending test results.

The first test pile was installed and a dynamic test of complete installation was carried out. Driving was completed at an installed length of 27.7 metres. Fig 1 shows the force and velocity curves recorded at the end of the drive. Pile capacity at this time was calculated at 118 tonnes.

The pile was restruck the following day. Fig 2 shows force and velocity curves at restrike. Pile capacity was now calculated at 173 tonnes.

The pile was statically load tested by maintained load method by jacking against kentledge and the test was initiated seven days after installation and carried out over a 36 hour period. Maximum load applied was 120 tonnes. Fig 3 shows the result of load.V.deflection for the pile and max. deflection at test load was 9.5mm. The pile was not tested to failure.

The pile was finally dynamically tested at re-strike 13 days after installation. Fig 4 shows force and velocity curves at restrike. Pile capacity was now calculated at 268 tonnes.

Note in comparison of Figs 1,2 and 4 the increasing divergence between the force and velocity curves indicating the "set up" effect of the soil on the pile shaft and therefore, the increasing pile capacity with time.

Following these results the second static load test was dispensed with and a further five piles were dynamically tested both at installation and 24 hours later. The increase in pile capacity in this 24 hour period varied from 31 tonnes to 81 tonnes. Minimum pile capacity 24 hours after installation was 154 tonnes.

2. <u>North West Lancashire</u> Twenty piles were installed as foundations for small factory unit. Site investigation generally indicated a thin capping layer of fill underlain by loose sands with the proposal that the piles should found in the underlying sandy clays. However, although a description of the soil was available no tests

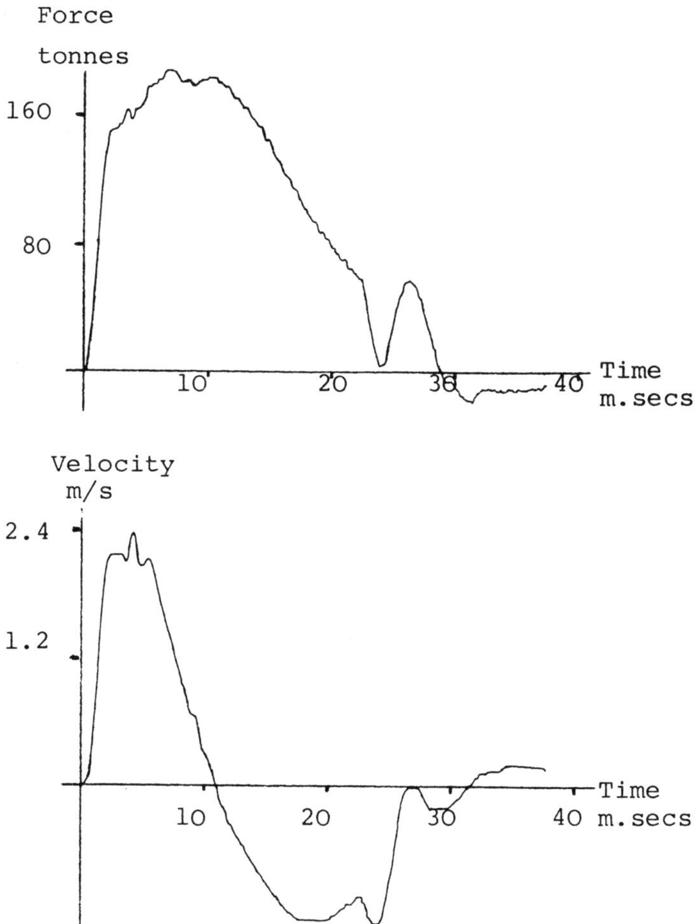

Fig 4. Force and velocity curves of test pile 13 days after installation.

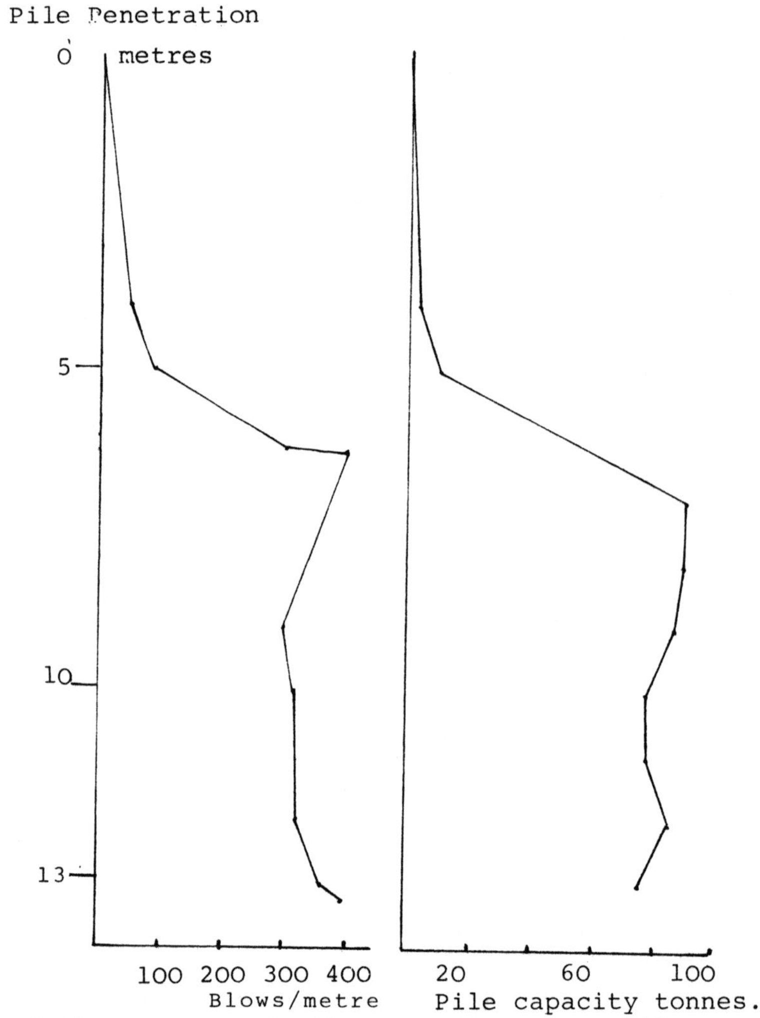

Fig 5 Blow count and pile capacity versus pile penetration North West Lancashire

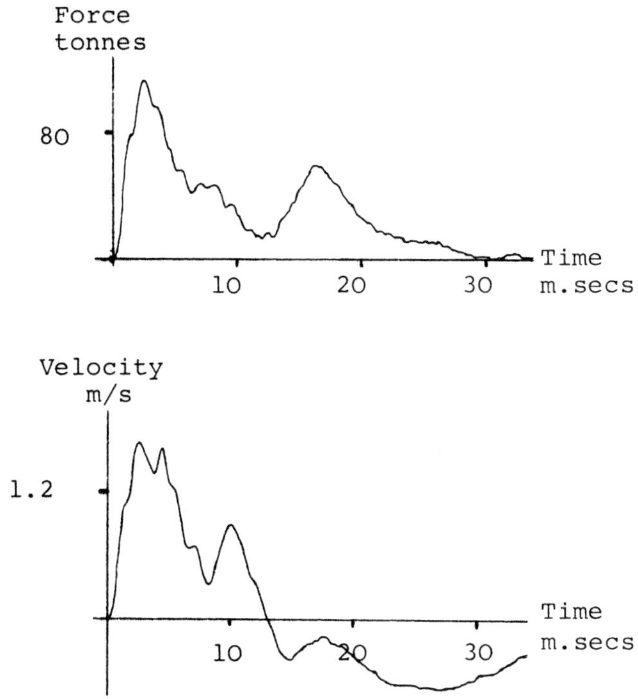

Fig 6 Force and velocity curves at 7 metres depth

had been carried out on the material to indicate physical properties.

Preliminary pile design was carried out by driving formulae to indicate a pile capacity sufficient to safely obtain a working load of 38 tonnes. Also the first pile installed was to be dynamically tested for its complete drive and bearing capacity with depth to be assessed.

The pile was driven to a depth of 13 metres without obtaining the required set (penetration resistance) as calculated by driving formula. On calculating pile capacity with depth from data obtained from dynamic measurement, it was found that the pile achieved the required capacity at a depth of 7 metres.

Fig 5 shows blow count and bearing capacity of this pile against depth.

Fig 6 shows the force and velocity curves for this pile at a depth of 7 metres and Fig 7 shows the force and velocity curves at a depth of 12 metres.

Note the curves of force and velocity at 7 metre and 12 metre depth are similar.

A further three piles were dynamically tested with similar results to the first.

3. <u>South Yorkshire</u> Twenty one piles were installed as foundations for shaft winder gantry. Site investigation report indicated up to 4m depth of fill underlain by dense becoming very dense sands. Piles were expected to found in the sand at a depth of 4.5 to 6 metres to sustain minimum ultimate load of 80 tonnes with minimum factor of safety of 2.

Programme was of prime importance and it was required that piling should be carried out in not more than three days. The specification called for one static load test to be carried out.

The alternative was put forward that two piles could be dynamically tested to obtain pile capacities during the period of the piling and save any extra time required in static load testing. Due to the congested nature of the site

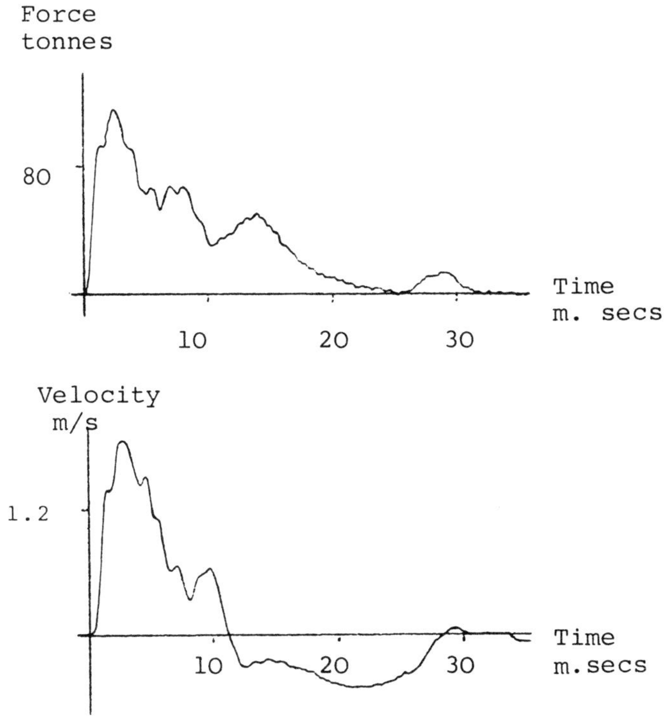

Fig 7 Force and velocity curves at 12 metres depth.

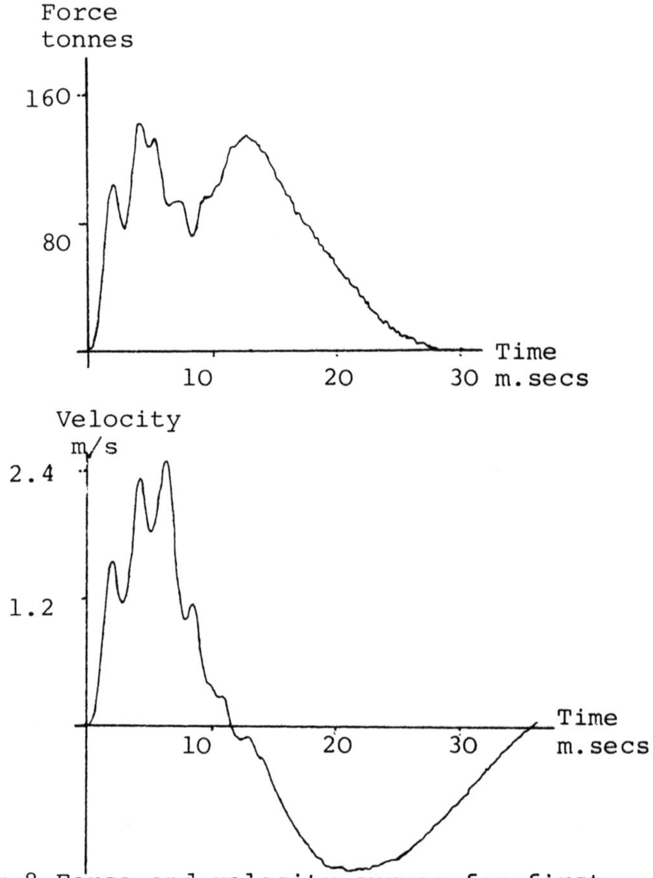

Fig 8 Force and velocity curves for first pile tested South Yorks.

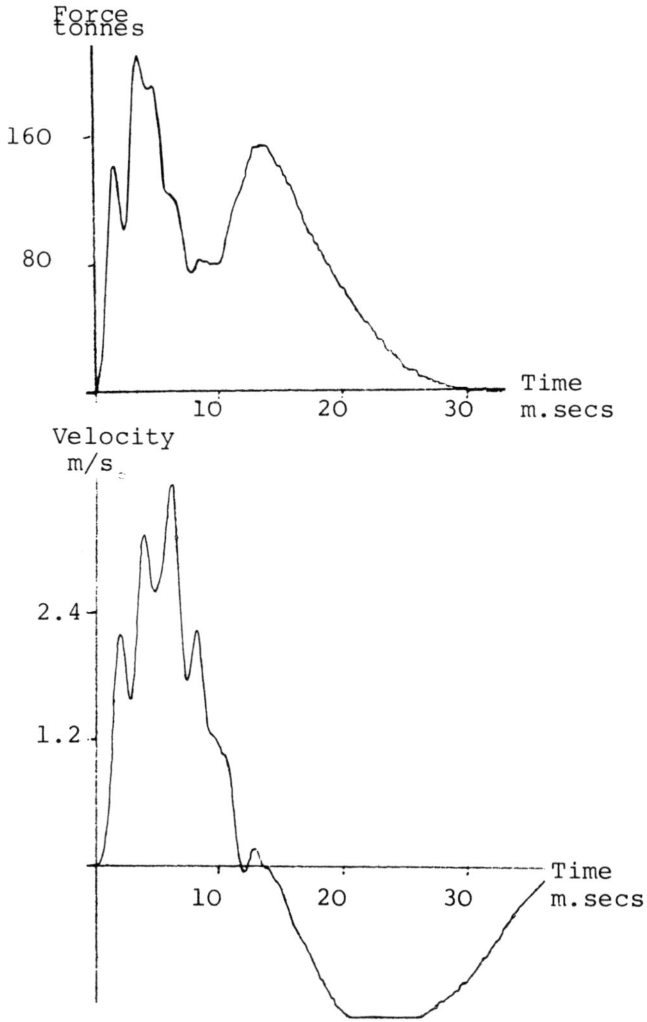

Fig 9 Force and velocity curves for second pile tested South Yorks.

it would not have been possible to carry out static load test during piling notwithstanding vibration effects of piling on test apparatus.

The first two piles installed were tested dynamically at time of installation throughout the driving.

Figs 8 and 9 show the force and velocity for each respectively at the end of the drive. Pile capacities were calculated at 162 and 159 tonnes respectively.

The piling works was completed including dynamic testing of two piles in two days.

References

1. Goble, G.G., Likins, G., Rausche, F.

   Bearing capacity of piles from dynamic measurements. Final Report No. OHIO-DOT-05-75 Department of Civil Engineering Case Western Reserve University, Ohio, March 1975.

2. Gravare, C.J., Goble, G.G., Rausche, F., Likins, G.,

   Pile Driving Construction Control by the Case Method Ground Engineering, March 1980.

3. Rausche, F., Goble, G.G.,

   Determination of pile damage by top measurements. A.S.T.M. Symposium Behaviour of Deep Foundations, Boston Massachusetts June 1978.

4. Gravare, C.J., Hermansson, I.,

   Practical experience of stress wave measurements. Seminar on Application of Stress Wave Theory on Piles. Stockholm June 1980.

5. Kightley, M.L.,

   Tapping into driving data. Contract Journal September 1980.

Discussion on Papers 1–3

Dr T.R. SHELLEY (*Paper 2*): Trials conducted at Ramsgate Harbour in February 1980 were part of a day of experiments on the site which was to be the terminal area for the new ferry service to Dunkirk. A number of novel techniques were tried, one of which was to use the Spectra-Physics rotating laser level.

Several errors were made in using this system. One was to mount the sensor at the paver tow-point instead of off the side-arm between the tow-point and screed so as to obtain feedback from the actual mat depth being laid. Another was to not mount the sensor higher so as to reduce the effect of convection currents. A further mistake was to expect the beam to give sufficiently accurate reference levels over a distance of 200-300 m.

The base was derived from colliery shale, which is black. All went fairly well until the sun came out. It was a cold day and although the base did not warm up much it was considerably warmer than the ambient air after the sun started shining. Convection currents were clearly visible and the laser beam scintillated so that the indicator on the paver was indicating 'Go up', 'Go down' and 'O.K.' in rapid succession. As a result, the operator of the paver returned to his usual method. He staked out a strong line, and placed a piece of timber on the screed so that he could match the mat depth to the string line.

At the end of Paper 2, I mentioned that better stability would result from using an infra-red system. However, this would not be worth doing because it would only halve scintillation. Achieving twice the stability would be well worthwhile, but equipment using infra-red light is not yet commercially available.

Sensors in highway and civil engineering. ICE, London, 1981, 35-38

DISCUSSION ON PAPERS 1-3

Mr P.B. STILL (*Principal Scientific Officer, TRRL*): I am involved in developing a system for measuring the profile of newly laid concrete surfaces and would like to apply it to newly laid asphalt surfaces. Is further information about the problems encountered with using lasers on asphalt surfaces available? Would a beam 3-4 m above the surface reduce these problems? Is there any indication as to what size deviations there are on the beam? Are they of the order of millimetres or tens of millimetres? Is there any way of getting over the problem by averaging the position over one or two seconds?

Dr SHELLEY: There is no detailed information on the magnitude of the instability. On the Ramsgate Harbour site, I would estimate the instability to be approximately 20 mm over 200 m or 300 m horizontal distance. The problem could be reduced by averaging. If personnel use a theodolite they do this automatically.

If a road is being laid or measured, the laser source should be off to one side of the road so that the beam passes over similar distances of the same kind of ground. This is better than having the source close to the road so that the beam approaches or recedes from the source. It has been recommended that the transmitter and receiver should be elevated to a height of 3 m.

Mr A.C.E. SANDBERG (*Senior Partner, Messrs Sandberg*): Are there further details available of the uses of reflectors, and examples of their uses? Why have British Rail apparently not taken the matter further following their experiments with track levelling?

Dr SHELLEY: A trial with the British Rail system took place in 1978. The laser was targeted on a plywood table 12 m ahead of the paver tow-point. The paver engine was running, but no material was laid. Considerable trouble was experienced with intermittent faults. Machine vibration was probably the cause and this has been found to create difficulties with any sensitive equipment mounted on a paver. The beam is projected from a laser on the machine, reflected from the base in front of it and picked up by a sensor mounted at a different height from the laser on the same machine. If the laser is aimed on to the base, the problem of convection currents from the roadway occurs.

The technique was developed by Britisn Rail for their track levelling. Blaw Knox has lost interest as a result of the malfunctions during the trails.

Dr S.F. BROWN (*Department of Civil Engineering, University of Nottingham*): With regard to asphalt paving, a 3 m high rod sitting on top would not seem feasible bearing in mind the accuracies needed and the problems of vibration. To what extent has this been thought through?

Was the instrumentation monitoring equipment (Paper 3) planned as a research exercise or is it suggested that these techniques could be used as a routine control arrangement for normal pile driving operations?

Dr SHELLEY: The problem of resonant frequencies is occasionally encountered when our present joint matching sensor is mounted on a long beam extending forward from the screed end. It is possible to select a beam length so that a severe vertical oscillation is excited by one of the machine vibration frequencies. The solution is to pull the beam in so that it extends to a shorter distance. The same kind of problem is likely to result from mounting a sensor on a 3 m pole. If it began to oscillate strongly, it would have to be retracted until it did not do so.

DR J.C. LEWIS (*Imperial College*): With regard to the Case method of modelling (Paper 3) tests on site are carried out on the pile as it is going in. There must be a resettlement effect. Has Mr Mure done any tests, say, 30 days after?

Mr J.W.M. VALLIS (*County Engineer, Warwickshire County Council*): Many engineers use cast in situ piles rather than solid piles. The usual method is to use a static load on a number of these. Is there any reason why this system should not be used to test cast in situ piles after they have been through their curing period? Is it substantially less in cost?

Mr E. BALLARD (*John Laing Ltd*): As technology advances, so normally the demands made on industry advance. If the laser type of instrumentation provides a more accurate tolerance band than that available from the old manual type systems for setting out and monitoring, are we already finding or might we in future find that the specification will narrow to match the increasing ability to meet the tighter tolerances and will we get a compounding situation? Is there any sign of this yet?

DR SHELLEY: In the UK a tolerance of 3 mm in 10 feet is required. If companies are able to improve on this and tolerances are made more severe, other companies will have to compete and meet the same standards.

Mr M. COTTELL (*County Surveyor, Northamptonshire County Council*): Specification tolerances are determined mainly through dialogue assessing what the client wants in terms of value for money and quality, and what the reasonable contractor and reasonable engineer can provide. Specification tolerances will get finer as the techniques allow them to be practically produced to a finer level. Ideally, the best possible road surface is ± 0.1 mm. Therefore, as techniques improve specificational tolerances will reduce and give better value and better quality to the client.

The techniques developed by Mr Twyning (Paper 1) and Dr Shelley (Paper 2) present the possibility of improving tolerances in a flat plane. What is not clear is how they achieve that improvement in tolerances in planes which are changing in three directions at once (the situation on most road alignments) where the plane is actually changing as one goes along the road - horizontally in two directions and also vertically.

Dr SHELLEY: This is under active study.

Mr J. SHREWSBURY (*Transport Planner, Halcrow Fox & Associates*): Is there any possibility of making a detector to measure the position rather than the presence of the laser beam?

4

Computer based traffic management systems for road tunnels

F. GERAGHTY, MIERE, Traffic Control and Communications Division, Department of Transport

SYNOPSIS. Traffic surveillance and plant monitoring in high speed road tunnels have, in general, being areas of particular difficulty. For safety and traffic management purposes various forms of vehicle control and regulation have been employed, necessitating the use of specialist staff. This has, on occasion, led to conflicting objectives being pursued, by the traffic police controlling the tunnel approaches and the tunnel controllers.

The advent of sophisticated low cost microcomputer based systems has led the Department of Transport to consider their use in tunnel management systems. The paper considers how such systems could be used to provide an integrated traffic and plant surveillance system for use by the traffic police without the need for specialist tunnel operating staff. The paper argues the cost benefits of such an approach and demonstrates the considerable technical and management enhancements to be had from such systems. The computor system hardware and proprietary software are areas of low development hazard since standard manufacturer supported equipment would be used. However, systems application software is an area of considerable development hazard where costs can, unless properly monitored, escalate to prohibitive levels. This applies also, but to a lesser degree, to the development of special transducers for monitoring and surveillance purposes. The paper considers how best to minimise these risk areas whilst providing the versatility and expansion capability inherent in programmable systems.

Sensors in highway and civil engineering. ICE, London, 1981, 39-45

The proposed use of microcomputer based systems for traffic surveillance, plant monitoring and general tunnel management provides a powerful impetus for integrating these, up to now, diverse tasks into one computer based tunnel management system, where real-time on-line control and monitoring will act to reduce operating costs, increase reliability and enhance safety.

INTRODUCTION

1. The management of traffic in road tunnels is significantly different from that of any other road system. Traffic engineers, in designing road tunnels, have long recognised this difference and have provided special facilities in both road layout and equipment systems. However, the control and regulation of tunnel traffic and the operation of tunnel plant equipment have, in the main, been seperate activities. To effect a satisfactory level of tunnel control and management has necessitated the employment of specialist operating staff. On occasion this has led to conflicting traffic management objectives being pursued by those responsible for the tunnel approach road network, and those responsible for the operation of the tunnel.

Microcomputers

2. A brief discussion on the technology of microcomputers will perhaps help in understanding the spectrum of application for these devices. It is no exaggeration to say that the advent of the microcomputer is one of the most important technological events of the second half of the 20th century. Its role in transportation systems will develop as traffic engineers learn to recognise its potential. However, a note of caution is in order since the misapplication of such systems can be both expensive and inappropriate.

3. The placement of microcomputers in the computer and logic systems league illustrates the bridging area which these systems occupy between hard wired logic, which is task specific, and programmable systems which are more universal in their application.

4. Microcomputer based systems are significantly cheaper than mini-computer or hard wired logic systems and their configuration capability more adaptable to a very wide

application spectrum. The systems can be wholly user transparent or system pre-eminent. This extraordinary flexibility arises from the component nature of microcomputers. In essence whilst mini-computers derive from a manufacturer developed design in fixed configurations, the microcomputer can be assembled using integrated circuit components from widely differing sources and technologies. It is true that manufacturers now offer total systems with complementary support chips to enhance and maximise the efficiency of their microprocessor, but it is not an essential requirement that all hardware need comes from a single source. Such flexibility is a double-edged sword where tales of cost effective applications are matched by very expensive mistakes. Detailed system analysis is essential when designing control networks using microcomputers.

5. The component nature of microcomputer systems was touched upon earlier and some expansion of this aspect should prove helpful in understanding. A microcomputer consists of a number of subsystems each located on one or more function specific integrated circuit chips (bugs) and mounted on a printed circuit board. (Alternatively several pcbs may be used). However, the packaging requirements for a microcomputer are significantly less than those of a minicomputer, as are the power requirements.

6. These sub systems are the microprocessor chip, memory chips, input/output interface chips and power supply module.

7. The functions of these chips are more or less self explanatory although their particular technologies and operation are extremely complex and beyond the scope of this paper.

8. Several manufacturers now offer microcomputers which are part of a larger family of machines and therefore offer to the systems designer a range of both hardware and software that is upwards compatible (ie programs written for the smaller machines in the family can be successfully mounted and run on the larger machines). The converse is also true but with certain important qualifications which are too complex for discussion in this paper.

9. In summary the microcomputers are low cost, compact and have a very wide application spectrum with very low power requirements. However, their employment in any control system application requires the most rigorous

attention to system design, documentation and maintenance philosphy.

Current Applications

10. The use of computers for traffic management in road tunnels is not new and is successfully employed both in Europe and elsewhere. However, these are by and large centralised control networks with a high degree of rigidity in their configuration. They are expensive to operate and maintain and whilst generally effective permit little automatic or manual reconfiguration in the event of outstation failure. This imposes a limitation on their efficiency since non-intelligent outstations require continuous access to the central processor for data transfer and outstation control. A great deal of central processor time is taken up with these data conveyance tasks and outstation status checks. The software overhead for this method of control is very high and failure rates both in outstation hardware and transmission systems,are a continual source of concern.

11. In addition existing tunnel computer systems are confined to a role as traffic management networks leaving both the environment and plant as seperate entities. The monitoring and control networks for the lighting system, the CO measurement system, the ventilation system, power supply system, closed circuit television system and emergency telephone system are dealt with seperately each having its own status display in a control room.

12. This seperation of plant and traffic monitoring is not necessary and in the authors opinion is undesirable, since it does not provide the tunnel and approach road management authorities with an integrated total systems status reporting network. The complexity of the mimic displays dealing with tunnel environment and plant status require highly skilled operatives to understand and control them. The costs of providing such systems is high and their maintenance somewhat costly. The flexibility of control afforded by such systems is limited and their expansion and/or reconfiguration capability low, without incurring a very high level of costs and associated development risk. In general therefore even the best of current systems can be demonstrated to be less than wholly satisfactory in that traffic monitoring, plant and environment control are seperate systems relying on the tunnel operating staff to integrate them for overall management.

Integrated Systems

13. Current technology makes it possible to provide a totally integrated traffic management, plant and environment control network for both the tunnel and its approach road network using low cost but sophisticated microcomputer based systems. In addition such systems would be wholly automatic in operation eliminating the need for specialist staff to be employed. The further elimination of costly cable networks and complex display mimics would further reduce capital costs, running costs and most importantly maintenance costs,whilst increasing reliability and thus safety.

14. The configuration of such systems would be on a distributed network with intelligent outstations at all traffic sensor, plant sensor and environment sensor clusters. To afford maximum reliability it would be essential to build the system as a flexible boundary reconfigurable network,where failure of any outstation would result in remaining outstations being aware (through intersystem checks) of adjacent outstation failure and automatically reconfiguring to cater for such failure. The network would then operate to a lower level algorithm whilst the failure persisted.

15. All outstations would be designed in a universal manner using common hardware. Station function would be defined by the firmware modules and each station equipped or sub-equipped as its function demanded. By employing universal hardware the economies of scale would operate to keep individual station costs low, maintenance would be improved and system down-time for repair very considerably reduced. Each outstation sensor cluster would permit the station to function independently within its own boundary. Boundary consensus and conflicts would be resolved by a second level network (using the same universal hardware with its own function firmware) whose task would be, control office servicing and communications control. This second level system would operate in a master-slave configuration with automatic change-over. Control office personnel would receive status reports on Visual Display Units for all critical parameters when thresholds were exceeded. They would have the facility of demand readout of any parameter. Hard copy would be provided of all status reports and operator instructions. Operators would be provided with facilities to override the system. Such overrides would be checked to prevent unsafe operation.

16. In any integrated system the role of closed circuit television (CCTV) cannot be ignored since such systems provide, for the operators, the only visual access to the situation prevailing in the tunnel and on its approaches. To date little effort has been made to integrate CCTV with the computer controlled monitoring and control system. Both systems have remained independent of each other, each providing its own output to the operator. Using micro-computers it is feasible to provide signals from the monitoring network to display status and location information re any incident on the CCTV monitors, and to automatically trigger video recorder devices to capture the occurrence of an incident for later analysis. In addition line and frame snatching techniques are being developed where real-time on-line data analysis of traffic behaviour is possible. Such developments would make continuous observation and detection along the whole length of the tunnel and its approach roads possible. Using inductive loop arrays, for vehicle monitoring purposes, however closely spaced, relies on transient measurement extrapolation to determine behaviour and detect incidents. Continuous monitoring, such as afforded by the video system, would markedly increase the reliable detection of incidents.

System Design

17. The design of systems, such as that briefly described above, poses some very difficult problems.

18. In general the procurement of hardware is relatively simple and straightforward. Attention to such matters as packaging, power requirements, transmission systems and interfacing apply with equal force to whatever kind of system is employed. However, it is essential to understand the architecture of the hardware and in particular its operating methods, speed and interfacing requirements before embarking on the design exercise. A proper detailed systems study must be undertaken prior to the production of function specifications. A complete understanding of how the network is required to operate in all its modes, including and most importantly its failure modes, must be available to the designer.

19. The area of greatest risk and difficulty is that of software where costs can escalate to prohibitive levels unless properly controlled and monitored. The design of software cannot be hurried nor can short cuts be taken. The key to successful software design lies in full and detailed systems analysis and design specifications

followed by comprehensive testing and debugging. It is essential that properly defined factory tests are designed to exercise the applications software and that all conceivable failure modes of the hardware are tested. It is most difficult to define and quantify system operation under failure of systems, sub systems and components. The environment in which traffic and plant monitoring equipment is required to operate is extremely hostile both electrically and physically and it is therefore mandatory that system tests should include, insofar as is possible, the environmental elements.

20. Fail safe routines, both hardware and software, will go some way towards ensuring safe system operation but may contribute towards an increased shut-down rate thus decreasing system reliability.

21. There are no easy solutions to these problems but careful and detailed systems design should go a long way towards containing them.

22. Intelligent systems are capable of a wide variety of behaviour and not all of it intended by the designer, especially in failure mode, so a high degree of system self checking is necessary to ensure safe, reliable and continuous operation.

CONCLUSIONS

23. This paper has but briefly sketched the systems possibilities for an integrated approach to tunnel management using microcomputers. The role that such devices could play is only limited by the imagination and willingness of designers to apply them.

24. Their power, flexibility and cost make it certain that their use will grow throughout the field of traffic systems engineering. This can only be for the benefit of the user and the nation.

25. An effective transport network is a vital asset to the health of the economy.

ACKNOWLEDGEMENTS

26. This paper is published by permission of the Controller of Research and Development and Director of The Transport and Road Research Laboratory.

27. Any opinions expressed in this paper are those of the author and do not necessarily represent the policy of the Department of Transport.

5

Sensor applications in road tunnels

G. FUDGER, MIStructE, Tunnels Engineering Branch,
Department of Transport

SYNOPSIS. Road tunnels, and their equipment, must be designed to provide an adequate level of safety for the user. A controlled environment is required together with measures to minimise the effects and potential consequences of vehicle breakdowns and traffic accidents. This paper describes the different aspects of a road tunnel environment, its control and the role of sensors in this. Generally tunnels must be illuminated during daylight according to exterior luminance which must be sensed. Vehicle emissions may affect health or visibility and those sensors used to monitor the tunnel atmosphere and other parameters in order to control mechanical ventilation are described. Monitoring of traffic, detection of incidents and data transmission are described in another paper. Consideration is given to the role sensors can play in monitoring emergency, telephone and fire points within the tunnel, plant status and serviceability, security of associated works, etc. This paper is written in the context of UK experience and an indication of the intended use of sensors in planned UK tunnels is also given.

Sensors in highway and civil engineering. ICE, London, 1981, 46-62

INTRODUCTION

1. Road tunnels, since they pass beneath obstacles, can and often do form very important links in the highway network. If a lower length limit of 150 metres is defined there are some 25 tunnels in the UK at the moment ranging from the Mersey Queensway, 3237m long, down to the Matthew Murray Tunnel in Leeds at 160m or Penmaenbach, 170m, on the A55 Trunk road in North Wales. Because of the advantages they can offer to the adjacent environment tunnels are often considered for particularly sensitive locations although their cost will only allow a few to be built.

2. The prime requirement for any tunnel is that it must provide adequate safety for the user whilst handling the traffic efficiently. The magnitude of this task will depend upon the size of the tunnel, its importance in the highway network and the availability or not of diversion routes. Over and above the need for public safety, the economic aspect, commercial or regional, could provide incentives to minimise any loss of use whatever the cause.

3. A tunnel is an enclosed space and adequate safety may be defined as the provision of a controlled environment suitable for the task, ie driving through it. The prevention of traffic delays and the rapid response to accidents, breakdowns, etc, in order to minimise their consequences is important and appropriate attention should be given to this during design and operation. In practice the control problem can be separated into two components; that of the traffic itself and that of the environment, ie illumination, ventilation, etc, although traffic management to improve vehicle flow will enhance the internal environment or at least reduce the ventilation demand (ref. 1).

4. The detection of traffic incidents, the acquisition of data, whether by loop detectors and/or television and data transmission is dealt with in another paper.

5. This paper will review, briefly, the environmental requirements, the equipment and its operation covering those parameters which might be measured and the sensors which could be used. This is in the context of the existing UK tunnels and some knowledge of European practice and in the light of design work to date of those planned for the Department's motorways and trunk roads.

LIGHTING

6. From the users point of view, a driver approaching a tunnel in daylight will see the well known 'black hole' effect. At a distance the interior of the tunnel will appear to be a relatively dark space surrounded by the portal structure and various proportions of the local topography or sky. As he enters his eyesight will have to adapt to the lower interior illumination. Lighting engineers endeavour to provide an economic level of illumination within a tunnel which is sufficient for the adapted eye to see the obstacles and to permit safe driving (ref. 2).

7. However any abrupt large change, especially downwards, in illumination is dangerous. The change from the potentially very high, yet variable, exterior level of luminance to the relatively low interior level is assisted by additional lighting fittings at the entrance. Typically there would be an entrance reinforcement zone capable of providing between 6.5% or 10% of the exterior luminance and several transition zones reducing this to the interior level. Similarly there may be exit reinforcement in a one-way tunnel, not as extensive as at the entrance, in order to cope with the more rapid upward change.

8. Sunlight illuminance could exceed 10,000 lux on a bright day. It has been suggested that the interior daytime lighting should be designed to 260 lux for a heavily trafficked urban tunnel down to 40 lux for a tunnel with little traffic at low speeds, depending upon the diffuse reflectance of the carriageway (Ref: 2 p 71). The entrance reinforcement with a dark carriageway could be designed for 1560 lux. At night-time a lower level with no entrance reinforcement is appropriate, roughly equivalent to good street lighting.

9. Because of the wide variation of exterior light levels during daylight, through the different seasons and weather conditions, the entrance reinforcement and the transition zone lighting fittings ought to be switchable in several stages. The smaller, more simple, installation may only require two stages, day and night, but the more complex or larger could justify four or five on the grounds of reduced energy consumption.

VENTILATION

10. All internal combustion engines emit pollutants. A wide range of substances have been found and most would be dangerous if allowed to accumulate (ref: 3). When ventilating road tunnels two potential hazards are considered; the need to dilute the toxic gases or the need to disperse the particulate matter obscuring visibility and mechanical ventilation is introduced when natural or vehicle induced air flows are insufficient (ref: 2).

11. A variety of toxic gases can be present but, with the current vehicle emission characteristics, carbon monoxide is sufficiently dominant that the others may ordinarily be neglected. The maximum allowable concentrations are related to the usual short time of exposure of a user. Typically a designer would take 250 parts per million (ppm) for the calculation of the maximum fresh air requirements and set the control switching at lower levels, eg if two stages are used then the first could come on at 100ppm and the second at 200ppm.

12. An alarm at 400ppm is usually provided to cover the need to close the tunnel the most likely cause being failure of the fans.

13. The second design case considers the problem of reduced visibility caused by the attenuation of light rays due mainly to diesel smoke, specifically carbon particles less than 2 micron equivalent diameter.

14. This is a fairly recent innovation and several properties have been used to measure the phenomenon (ref: 4). The problem is that visual range can not be measured directly and none of the physical parameters are necessarily representitive of what an observer actually sees. This is not the place for an exhaustive review of the problem but it should be noted that a variety of methods of measurement have been used in the past with instruments giving an output signal in a variety of units eg gravimetric mg/m^3, Owen Fog Shade scale, % transmission and others.

15. Generally designers favour the extinction coefficient $K\ m^{-1}$, as defined by:-

$$T = e^{-Kd}$$

where T = % transmission of light

d = length of light path m.

16. Lately it has been proposed that apparent contrast of an object at a distance would be more representative of the visual task (ref. 5).

17. However whilst those concerned with the establishment and enforcement of standards must have an appropriate unit of measurement, the individual tunnel operator may seek simplicity in an instrument calibrated in arbitary terms such as the reaction of a set of observers during a series of test runs. This has not always had the desired effect. Ventilation levels based upon seemingly successful tests have produced considerable public complaint in one case at least (ref. 6).

18. The fresh air requirement to give adequate visibility is usually based on a maximum extinction coefficient of $6 \times 10^{-3}\ m^{-1}$ for traffic at 40 km hr^{-1} or faster and $9 \times 10^{-3}\ m^{-1}$ at 10 km hr^{-1}. Thus slower moving traffic, whilst producing more smoke, in fact does not need as great a visibility distance as the faster. A further upper limit based upon a 'comfort' or 'hygene' factor has been suggested at $12 \times 10^{-3}\ m^{-1}$ (ref. 2).

19. Therefore the primary control parameters of the fans may be the carbon monoxide concentration or the extinction coefficient. If mainly commuter traffic is involved either a manual or an automatic time programme based on traffic data could be used and the pollutants would only be measured for added security.

20. Some ventilation systems move the air in ducts connected to externally located fans and the volume flow in the duct may be a control parameter. Yet again the air flow in the traffic space whether velocity and/or direction may be of consequence and this might be measured although the turbulence due to passing vehicles will cause problems. Air flow measurements also confirm that the fans are working as intended.

21. Generally one would only measure pollutants within the tunnel although where the vitiated air is discharged in a sensitive area a further measurement may be imposed, eg at the Elbe tunnel, carbon monoxide is measured downwind of an exhaust shaft in the direction of a hospital giving an alarm if a set level is exceeded (Ref. 7).

22. A significant feature of a road tunnel is that it will contain a considerable mass of air and the fans themselves could have considerable inertia or equivalent effect such that a control command can only be fully implemented in a matter of two or three or even more minutes. It could also contain a considerable number of vehicles. Thus a saving feature in respect of the instrument design and specification is that one is principally interested in the average situation. An over-rapid response is not required since the ventilation system would not be able to react quickly. Also over frequent control changes would greatly increase wear and tear as well as the cost of energy. Whether the measured values are averaged at the instrument or the controller is appropriately programmed will depend upon the designer's approach to the control logic.

PUMPING

23. Roads in tunnels often have a sag point and water or other liquids, whether run-off, ground water infiltration or spillage, will have to be dealt with. If gravity drainage is not feasible then sumps and pumps will be required. These ought not to present any special problem of instrumentation or control, although the possibility of inflammable vapours being present giving a risk of fires or explosions must not be ignored. If electrical equipment must be located in such a space then it should be flameproofed or instrinsically safe.

EMERGENCY TELEPHONES AND FIRE POINTS

24. For various reasons road tunnels have fewer accidents than the equivalent length of average road in the open air. However an incident below ground could have potentially more serious consequences. It is usual to provide emergency telephones and fire points at quite close centres in the tunnel such that the user can raise an alarm quickly and has some means of dealing with the smaller fire at least prior to the arrival of the Fire Brigade. The telephone system would connect with the

Police traffic control room, usually remote from the tunnel where there would be some form of synoptic display showing the location of any telephone in use. Thus it could be said that the system is a sensor indicating the location of someone calling for assistance. Loop detectors and closed circuit television will enhance the detection of or reaction to incidents. Push-button alarms and switched cabinet doors or extinguisher brackets have also been used.

25. Fire detectors are occasionally installed in the tunnel itself. However there are operational limitations and the type of detector must be carefully chosen. Measurement of carbon monoxide and diesel smoke for ventilation control provides an added safety feature in that smoke and fumes are the more usual and more dangerous products of virtually any vehicle fire. A rapid simultaneous rise in both parameters probably provides a more certain indication of a fire.

OTHER SERVICES

26. The equipment in a tunnel will obviously need an electrical power supply involving high voltage feeders, transformers, low voltage switchgear, metering, plant control systems, etc. In addition to measurements as described above a fault monitoring system will be necessary. This equipment is usually housed in a separate building. There may also be a control room and the usual offices for staff. Even if meant to operate unattended certain facilities have to be provided. All this space may need protection from fire, intruders etc.

TUNNEL OPERATIONAL ARRANGEMENTS

27. The extent of the need for sensors in any particular tunnel will depend upon its importance, basically its length and volume of traffic carried and upon the type of operational arrangement adopted. The earlier tunnels were operated manually and still are to a degree. However the cost of providing 24-hour per day attendance by competent personnel is such that automatic unattended control for many average length tunnels must be the objective. The commuted capital cost of just a single person for three shift attendance plus supervision and administration for operational surveillance alone is such

that close consideration of reliable automatic controls is justifiable.

SENSORS

28. Lighting Control: The usual form of light sensor used for the switching control of entrance and exit reinforcement zones is a conventional photo-reactive device comprising one or more photoelectric cells (ref 8). These provide a simple on/off signal at a pre-determined level responding to the general exterior light level or to reflected light from the portal, the carriageway or other appropriate surface. The actual location of such a sensor is often influenced by the surroundings (ref 9). In any case the actual property measured is only part of a complex variable and some on-site experimentation is usually necessary.

29. A recent application of a videcon tube is of particular interest. A TV camera is directed at a tunnel portal taking in approximately the field of view that a driver would see, albeit stationary and at one distance only. The output from the videcon tube represents the integration of reflected light from this field of view and is varied by the changes in luminance or brightness of the scene thus providing the control signal.

30. The change from the day-time to night regime may have back-up control provided by a time-switch or closk with a standard solar dial to allow for changes in the time of sunrise or sunset throughout the year.

31. This application of light sensors is of particular interest in that the detection of changes in levels of luminance during bright daylight as well as those at dawn or dusk is involved whereas only the latter is usual in outdoor lighting.

32. Thus the lighting control functions are exterior luminance or the illuminance of an appropriate surface such as the portal or carriageway. However the actual switching levels required of any sensor will depend so much upon the specific nature of each installation that generalization in a paper of this nature is not possible.

VENTILATION CONTROL

33. If a tunnel is mechanically ventilated the carbon monoxide concentration will usually be measured, and also probably the extinction coefficient, at the point or points of maximum pollution within the tunnel. The actual location will depend upon the type of ventilation system, the available space and sometimes access. If actually in the traffic space then durability and reliability is of extreme importance since frequent visits for maintenance and recalibration are most undesirable.

34. The nature of the suspended particles requires that the extinction coefficient be measured as close to the drivers line of sight as possible. These particles, mainly carbon with small quantities of silica and rubber but with aromatic gums and unburnt hydrocarbons and oil vapour, adhere readily to any surface and this poses problems in trying to pipe samples to remote locations.

35. Carbon monoxide meters can be sited in a more protected environment and supplied with continuous samples. At least one manufacturer offers an automatically switchable manifold so that samples piped from several points may be measured sequentially by a single meter. An adequate sample flow rate should be employed since a low velocity in a long tube will impose a time delay.

36. Tunnel operators are interested in measuring the carbon monoxide concentration through a range from zero up to 400 or 500 parts per million and devices, reasonably well proven in the tunnel environment, are available from several manufacturers.

37. Of the possible analytical principles only two are in general use; catalytic oxidation and measurement of the heat produced or absorption of infra-red light. For a tunnel operator there may be little to choose between the better instruments of either type. One drawback of catalytic oxidation is that other combustible vapours will give a positive reading. In addition the sample has to be conditioned, mainly filtered and dried, in order to preserve the catalyst which needs replacing from time to time. All gases absorb infra-red light at specific wavelengths and manufacturers of the second type of instrument select the one least affected by other gases or vapours likely to be present.

If adequately protected in a robust construction the infra-red type is felt to have potentially fewer maintenance problems. Most manufacturers also offer automatic in-service recalibration devices as extras. A review of the available instrumentation is given in reference 10.

38. Any instrument will require specialist attention and recalibration from time to time. The problems of access, particularly to the interior of a motorway tunnel, must not be ignored and instruments should be capable of running unchecked for at least the time between planned maintenance sessions.

39. Measurement of visibility by one of its physical properties is only a partial representation of a complex problem. Early instruments were derived from chimney emission monitors converted to measure tunnel smoke densities some several orders of magnitude lower. More recently instruments developed to measure fog, particularly on airfields, have become available which have the required sensitivity.

40 Several different principles have been used in the past but only those based upon an optical parameter find favour today. The available instruments either involve the measurement of light lost in transmission over a fixed path or scattered by particles through a given angle or the loss of apparent contrast of a target when viewed at a distance.

41. The simpler transmission instruments, usually called transmissometers, have a standard incandescent lamp projecting a known light flux over a long path, typically 30m, measuring the received light by a photoelectric cell. Lamp ageing, cell drift, dirtying of optics and the impossibility of in-tunnel recalibration severely limits this type. Another version using a 150m folded light path has a chopper device permitting the comparison of received light with a reference beam transmitted internally which overcomes some of these limitations although flapping ropes and tarpaulins on passing vehicles frequently displace either the projector or reflector sufficient to disable the instrument.

42. Light passing through a turbid medium is scattered as well as absorbed and measurement of this is possible. Scatter instruments are compact, which is advantageous, but only give a point reading thus the averaging of several successive readings or a reliable air flow in the tunnel is required in order to positively represent the drivers' problems. The relationship between absorbed and scattered light depends upon the nature of the particle and the constancy of this for small carbon particles, ie black diesel smoke, in the presence of the other possible particles in a tunnel needs to be established. The relationship is effectively linear in a pure smoke and appears to be sufficiently so in some tunnels as to be adequate for ventilation control and the limitation is only apparent if more absolute data are necessary.

43. Visibility, ie seeing a target at a distance, is a complex problem and the most recent work indicates that "humans perceive visual air quality changes in direct proportion to apparent target contrast" (ref 5). Freibauer et al (ref 11) have demonstrated the significance, in a tunnel, of carriageway luminance which reaffirms the point. Marsault and Sorlie (ref 12) used a prototype device in their trials which showed promise although with problems.

44. Airfield visibility and the fog on motorway research programmes have produced some short base transmissometers having a structural frame to overcome the displacement problem and a discharge tube light source such as the gallium arsenide emitter (ref 13). They promise long term stability and a lower maintenance commitment which if borne out in practice will be welcomed by tunnel operators.

45. It may be necessary to measure air flow in some ventilation systems in order to verify the proper working of the fans. Measurement of volume flow in a duct follows standard practice; ie differential pressure drop across an orifice plate could be used although a pilot tube properly located is usually sufficiently accurate.

46. The need to measure air flow, specifically velocity and direction, in the traffic space will only occur in those tunnels where the traffic or wind induced movements are significant. This will be the case in particularly exposed locations through hills or mountains especially if the tunnel has a single tube carrying two-way traffic

where it may be necessary or economic to allow for these influences in the ventilation control.

Such sensors are generally of standard design being propellor or rotating cup anemometers and vanes although hot wire anemometers have been used experimentally and laser devices are now available. However the method of installation must take the turbulence caused by the traffic into account. Sometimes it may be possible to locate the sensors externally and calibrate them with respect to internal air movements. In either case the quality of the data could be limited but providing this is recognised suffiecient for the purpose can normally be acquired.

PUMPING

47. Where necessary the drainage system will discharge into a sump or sumps. Pumps will be required to lift the water to the outfall and these will be controlled in the usual manner by electrode level detectors or pressure transducers. Inflammable gases or vapours could accumulate in these sumps or indeed any similar enclosed space and it is usually advisable to install some form of hydrocarbon vapour detector set at something below the lower explosive limit which for petrol vapour in air is 1.3%. Occasionally it may be necessary for men to enter these spaces and an additional danger could be that of a reduced oxygen concentration. Whilst a permanently fixed oxygen meter may not be justified a portable instrument certainly can.

PLANT STATUS AND SERVICEABILITY

48. A road tunnel could contain considerable quantities of plant and equipment which would generally be distributed throughout its length. In addition to environmental monitoring, data signals from plant mounted transducers will indicate the plant status, ie working or not, and serviceability ie, fault condition. Even if the tunnel is permanently manned by technical staff much of the equipment will be remote from the control centre making it impossible for it to be directly over-seen. If the tunnel is operated automatically and largely unattended it will still be necessary, in fact more essential, to inform the Police of any condition which might have a potentially dangerous effect on traffic. For example, failure of the power supply or of all or

of all or part of the lighting or ventilation system could mean that the tunnel should be closed or that traffic restrictions should be imposed.

49. One approach is to classify all signals into emergency, ie do something now, or routine ie can wait until morning. The emergency signals would be dispatched immediately and the routine sent once per shift or only upon interrogation of the recording equipment which might be anything from a paper chart to a computer memory store.

50. A well designed tunnel will incorporate several devices or arrangements essential for good management. "Hours-run" meters on all plant enable reliability to be monitored and thus the proper planning of maintenance. In order to limit maximum current demand, cascade or sequential starting is essential for all groups of electrical equipment which operate simultaneously such as lights or booster fans. Duty rotation or load sharing between several items of equipment which make up one function eg, lights, fans or pumps is recommended good practice. All these depend upon the appropriate sensors and whilst it may be common knowledge to the process control engineer the highway engineer involved in a tunnel ought to be aware of the possibilities.

CURRENT UK PRACTICE

51. This will be dealt with at length in a TRRL report, 'The operation of road tunnels', in course of preparation. Generally the sensor applications in UK road tunnels follow from their method of operation, level of traffic and experience. The toll tunnels which have a common policy of very active intervention by traffic patrols, CCTV surveillance, hazardous material inspection and escorts, generally have high levels of traffic. Thus the environmental requirements are reasonably predictable such that the duty controller who is present to oversee the other duties can operate most plants remotely to a time-table or on feed back from the mobile patrols or by observation of the data displayed on the control panel.

52. Lighting works automatically according to sensor output. Ventilation is normally manually operated to meet traffic demands although standby fans come into action

automatically should the duty fan sensors indicate faults such as high bearing oil temperature or electrical failure. Air-flow in thousand of cubic feet per minute is measured and displayed. Carbon monoxide is measured by catalytic oxidation and visibility by transmission loss over 30m. The latter sensors require considerable maintenance. The ventilation systems have been investigated here and at Blackwall and the effects of short-comings in the sensing and measurement of pollutants upon the ventilation control noted (ref 14). Blackwall uses a smoke meter which measures the increase in pressure diffential across a filter paper tape when dirt is deposited. It is of limited effectiveness in automatic control and of course measured all suspended particles including those above 10 micron which have much less influence on visibility than those about 1 micron.

53. Rotherhithe Tunnel was closed for refurbishment on 4 August 1980. Up to that date the lighting which had little entrance reinforcement stayed on permanently. The fans were switched on manually to a time table regardless of traffic. On completion of the current work which includes up rating of the lighting and ventilation an automatic control system will operate.

54. The Clyde Tunnel operates in general as the toll tunnels but to a lesser intensity of manual supervision and intervention.

55. Other UK tunnels are shorter and do not have the steep gradients of the under river tunnels. Whilst all would have daytime lighting with entrance reinforcement few below 400m in length have mechanical ventilation.

56. Catalytic oxidation is used to measure carbon monoxide at Birmingham whilst Heathrow uses infra red absorption.

57. Generally the sensors provide a signal which is used to actuate an electromagnetic contractor in order to control the specific item of plant.

NEW WORKS

58. In the UK the Lewes tunnel has just been completed. This is a single tube, 420m long, with one traffic lane in each direction. The lighting controls are conventional.

The ventilation is of the longitudinal type with single speed reversible booster fans hung in the the arch and controlled, in direction and operation, by a time switch to suit the predicted commuter flows. An automatic interlock blocks the start signal if the wind is sufficient to ventilate the tunnel.
Carbon monoxide is measured at two locations within the tunnel and wind speed and direction on top of the switch room building outside. Emergency telephones link the tunnel to the East Sussex County Police operations room who can remotely operate "tunnel closed" secret signs.

59. Work has just started on the construction of the Holmsdale Tunnel on the A10 to M11 section of the M25. Lighting control will be by photocells in 5 stages. The longitudinal ventilation system by booster fans will be controlled by carbon monoxide or diesel smoke measurements. Because of the expected volume of traffic and the distance of the Motorway Police control room from the tunnel the problem of data transmission is important. One effect is that sensors with digital outputs are preferred making the necessary electronic processing and supervision of the message transmission easier. The Bell Common tunnel, further east, will be similar although shorter.

CONCLUSION

60. The environmental, operational and safety aspects of road tunnels have been reviewed and the application of sensors broadly indicated in order that highway engineers may be aware of the need to consider their provision and in service use when contemplating or planning road tunnels.

61. It is of interest to note that a considerable number of tunnels have been built in since 1960; both in the UK and in Europe. Much attention and research has been given to their design and construction since this gave the greatest scope for capital cost savings less to the problems of equipment control. The equipment in the earlier tunnels is coming to the end of its working life, eg the lighting system in both the Clyde and the Crindau tunnels has been renewed recently. One can anticipate further work of a similar nature. In addition increased running costs will mean that more attention will be given to operation and maintenance. In either case the opportunity to re-examine the application of sensors to a

particular tunnel must be taken and it is expected that modern solid state devices and semi-conductor technology, both for sensors and controls, will play a greater role than hitherto.

ACKNOWLEDGEMENTS

62. This paper is published by permission of the Deputy Secretary, Roads and Local Transportation Group of the Department of Transport.

Any opinions expressed in this paper are those of the author and do not necessarily represent the policy of the Department of Transport.

References

1. FOOTE R S. Research for Optimal Ventilation at the Holland and Lincoln Tunnels. Proceedings of the 1st International Symposium on Aerodynamics and Ventilation of Vehicle Tunnels (1st ISAVVT). BHRA Fluid Engineering, Cranfield, 1973, Paper B3.

2. Technical Committee for Road Tunnels. Report to XVI Congress, Vienna 1979. Permanent International Association of Road Congresses (PIARC), 43 Avenue Du President-Wilson, Paris 16.

3. PURSALL B R. The Aerodynamics and Ventilation of Vehicle Tunnels. A state of the art review and bibliography. BHRA Cranfield 1976.

4. Technical Committee for Road Tunnels. Report to XV Congress, Mexico City 1975. PIARC, p29 et seq.

5. MALM W. Considerations in the Measurement of Visibility. Journal of the Air Pollution Control Association. Vol. 29 No. 10, October 1979 pp 1042-1051 and correspondence from CHARLSON R J ibid Vol. 30, No. 2 February 1980 pp 114-115.

6. FOOTE R S. Port Authority of New York and New Jersey. Private communication.

7. HERZKE K. Ventilation System of the New Elbe Tunnel in Hamburg. Proc 2nd ISAVVT, BHRA, 1976, paper B2.

8. SCHREUDER D A. Measurements, Public Lighting, ed J B DE BOER. Phillips Technical Library, Eindhoven 1967, pp 433 et sq.

9. HUGHES O G AND ROBERTS A. Re-lighting of the Penmaenbach Tunnel. Public Lighting, 45, 2, June 1980, 75.

10. CIRIA. A review of instruments for gas and dust monitoring underground. Construction Industry Research and Information Association. Report 80 October 1978.

11. FREIBAUER F, PISCHINGER R AND PUCHER K. Austrian method for calculating the fresh air demand in road tunnels. Proc 3rd ISAVVT, BHRA 1979, paper F2.

12. MARSAULT J P AND SORLIE P. Methods of Measurement for the control of road tunnel pollution caused by smoke from vehicles. Proc 3rd ISAVVT, BHRA, 1979, paper F3.

13. DOUGLAS H A, JEFFREY D J AND JEZZARD F. The measurement of fog on motorways. Meteorological Magazine Vol. 107, 1978 pp 242-249.

14. PURSALL B R AND SWANN C D. Research into the control of vehicular tunnel ventilation systems. Proc 1st ISAVVT, BHRA, 1973, paper B2.

6

The loop detector and its applications

R. K. DULEY, MIERE, MBIM, Redland Automation Ltd

SYNOPSIS. This paper briefly reviews the history of the inductive loop vehicle detector, some of its applications, the principles of its operation, some loop designs and installation criteria and the factors to be considered when designing or using the detector itself.

HISTORY

1. During the early fifties in the U.S.A., the proliferation of the motor car led to the introduction of car parks controlled by automatic barriers. It was soon found, however, that the old, direct contact vehicle detectors were unreliable and other devices such as light beams were unsuitable since they were activated by pedestrians as well as cars. A new sensor was thus needed which would be reliable, need no maintenance, give an output while the vehicle was present and not detect pedestrians. The first inductive loop vehicle detector was built and found to meet all the above requirements. So, the Sarasota Engineering Company who had made this prototype detector began production of the world's first loop detector.

2. During the following ten years automatic car parks spread throughout the western world and with them went the loop detector. It was during this period that the loop detector came to the notice of traffic engineers. Some traffic signals were already vehicle actuated, using pneumatic tube detectors, but the vast majority were still fixed time with the exception of those in the United Kingdom where a policy of vehicle actuation had been adopted 30 years before. It was soon realised that the loop detector could be used to indicate the formation of queues due to its presence detection capability and it was installed to call special movements such as early cut-off stages. By then, traffic densities had increased to such a level that

Sensors in highway and civil engineering. ICE, London, 1981, 63-72

it was often impractical to maintain pneumatic detectors in a working condition. The time was ripe for the loop detector to take over as <u>the</u> sensor for traffic signals and once again the United Kingdom led the world when the Ministry of Transport, as it then was, decided that all vehicle actuation should be converted to loop detectors. Today you will find the loop detector being used as the sensor for calling right of way, extending right of way, measuring vehicle speeds on high speed roads, indicating queues and calling special movements. In urban traffic control schemes you will also find the loop detector sending flow occupancy and queue information back to the central computer for future plan generation and now, in Glasgow and Coventry, as the input sensor for S.C.O.O.T.

3. During the energy conscious seventies the most rapidly growing application for this sensor was the automatic control of factory doors for vehicular traffic to minimise heat loss. But now it is the turn of the traffic surveillance systems to feel the benefit of the loop detector. Permanent and semi-permanent traffic counting sites have used loop detectors for some years now and more recent applications include the measurement and analysis of vehicle speeds, vehicle lengths and traffic headways.

4. <u>Principles of operation</u>. A loop of wire has an inductance which is governed by its perimeter, its width, the effective radius of its wire cross-section, the number of turns and the permeability of the medium around it. When this loop is buried in a sound road surface all of the above are constant and thus the loop possesses a stable inductance. If an alternating current is passed through this loop, alternating electric and magnetic fields are set up around it. Should a vehicle, or any object containing electrically conductive material, enter this field, eddy currents will be induced into it which will, in turn, produce magnetic and electric fields that couple with the loop, reducing its inductance. Road vehicles often contain considerable quantities of ferrous material which will increase the net permeability around the loop and hence its inductance but, provided the loop is correctly designed, the eddy current effect will dominate and an overall reduction in inductance will be experienced. This reduction in inductance can be detected by various means which will be discussed later.

5. The strength of signal given by a vehicle is a function of the amount of magnetic flux from the loop that links with the vehicle, and the conductivity and permeability of the

vehicle material. Thus the change in inductance can vary from 4% for a normal saloon car over a one lane rectangular loop to 0.02% for a bicycle over a six lane chevron loop.

6. The loop is connected to the detector electronics by means of a twisted pair of wires called a feeder. The feeder wires are twisted to maintain them in close proximity thus reducing the magnetic and electric fields around them and minimising their inductance. This has three advantages:

a) The containment of the fields reduce coupling and crosstalk with other adjacent feeders.

b) Since the fields are limited they do not couple with vehicles passing over the feeder, thus avoiding unwanted detections.

c) The dilution effect of the fixed feeder inductance on the signal produced by a vehicle is minimised.

7. Loop design. It will be readily appreciated that optimum performance can only be achieved by ensuring that the loop is correctly designed for the application in hand. A few common loop types are illustrated in the following figures but it is always advisable to consult the detector manufacturer before installation. Mistakes can be expensive!

a) Closing loop for automatic barrier. (see Fig.1). One of the smallest loops in common use, usually consisting of 3 turns. The loop dimension 'X' is a function of the lane width and would typically be 2 metres in a 3 metre lane. The 0.8 metres in direction of travel has been kept as small as possible so that the barrier will close as each vehicle leaves the loop and before a subsequent vehicle enters. This dimension is only suitable for car parks which are limited to private cars. N.B. Motorcycles should not use automatic barrier car parks.

b) Loops for automatic factory door control. (see Fig.2) These doors would normally be of 2 turns and connected in series so that the currents in them would be in phase. This reduces the magnetic field strength between the loops and creates a 'dead' zone halfway between them where a metal door is free to slide without causing unwanted detection. Dimension 'X' is, as before, dependent on the door width and 'Y' is a compromise between the need to maximise the distance of the loop from the door and the need to ensure that the shortest vehicle will enter the second loop before leaving the first. Loop 2 should be an exact mirror image of loop 1 and the

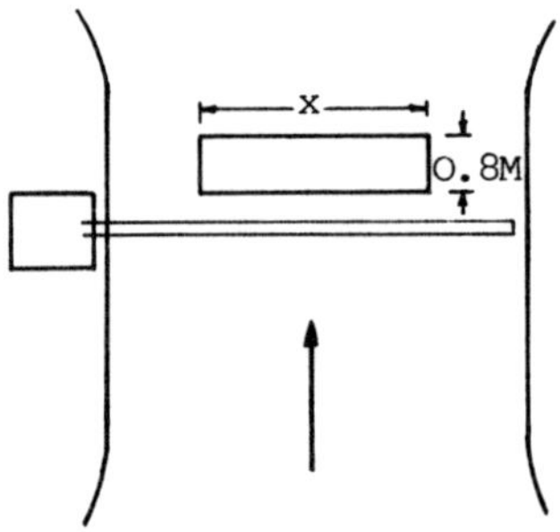

Fig.1. Closing loop for automatic barrier

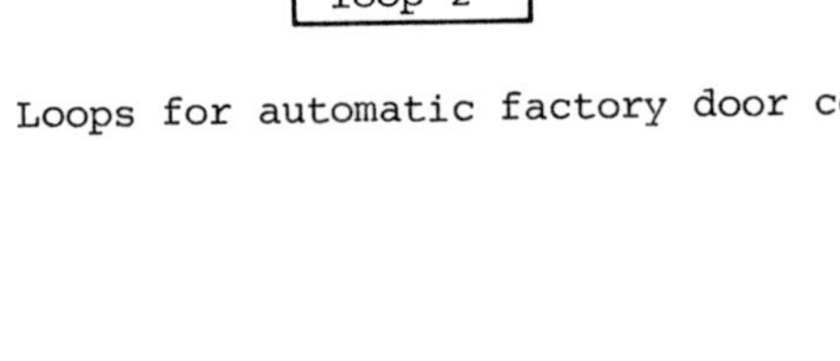

Fig.2. Loops for automatic factory door control

2 metre dimension is chosen to give continuous detection from a high chassis vehicle.

c) Bicycle detection in traffic control. (see Fig.3). In order that reliable detection of bicycles may be achieved it is necessary to ensure that they cross the loop wire at an angle of about 45^0, thus causing them to cut more lines of magnetic flux. The optimum loop dimension in direction of travel is 1.1 metres, 'X', as usual, depends on lane width and it is normal to use 2 turns. If this type of loop needs to cover more than one lane it should turn through 90^0 at the lane boundary to form a chevron so that the mean detection point in each lane is the same distance from the stop line.

d) Speed measurement. (see Fig.4.). The most important factor when installing a pair of loops for speed measurement is to ensure that they are in all respects identical. This produces the same profile for each of the zones of detection so that a given vehicle passing over the system will penetrate each loop by the same amount before detection is given. The speed measuring base is not critical as long as it is accurately known but 3 metres leading edge to leading edge is typical. These loops would normally be 2 turns and 2 metres in direction of travel.

8. The above loop configurations are just four examples of common applications, there are of course many more. For example:

- Traffic counting
- Queue detection
- Directional sensing
- Critical gap sensing
- Tram detection

If you have a requirement to detect the presence or passage of vehicles it is well worth considering the inductive loop.

9. Loop installation. The cost of installing the loop and feeder is often greater than the purchase price of the detector itself so you only want to do it once. The emphasis should therefore be on high installation standards with the work preferably being supervised by a member of your staff. Once a loop is buried it is difficult to tell where the corners (metaphorically) and the wire installation (actually) have been cut, until it fails a year or so later.

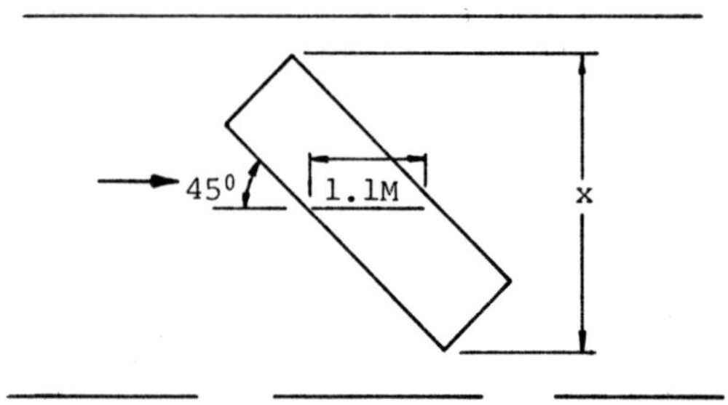

Fig.3. Bicycle detection in traffic control

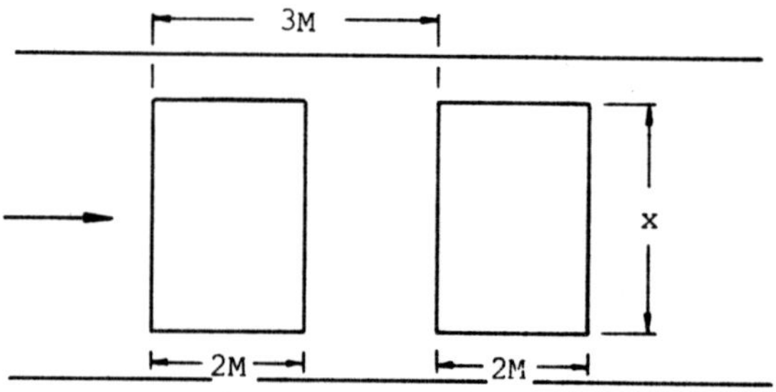

Fig.4. Speed measurement

10. Modern loop detectors are designed to track out a considerable amount of environmental drift, 15% in the case of some Redland units, but this can only delay the evil day when a damaged loop will cause the detector to need re-tuning. A good measure of loop damage is its insulation resistance to earth which should be measured at 500 V and should read in excess of 10 MΩ. Our detectors will operate satisfactorily with as little as 500 Ω across the loop terminals, provided that the value remains stable. But if you measure less than 10 MΩ to earth the loop or feeder is damaged and with time and changing climatic conditions the loop impedance will change sufficiently to cause the detector to need re-tuning.

11. The measured mean time between failures of our detectors is of the order of 40 years. Where high loop installation standards have been employed in the U.K., M.T.B.F., figures of over 11 years have been achieved for the loop/detector combination and overseas, where fewer utilities are buried under the road surface, even this has been exceeded. In other areas, when installation costs have been cut, figures of only 2 or 3 years have been observed. The message is, "Spend a little more to make a quality installation and save on the total cost of ownership". As an added benefit your system will suffer less down time.

12. Some hints on good loop installation technique follow:

- Ensure that slots are cut to the D.Tp. recommended depth (see ref.1) and that they are free of sharp objects and corners.
- Slots should be clean, dry and dust-free before the loop wire and backfill are installed.
- Backfill material should be hard enough to resist sharp objects being driven into it by traffic, but flexible enough to move with the road surface.
- Use loop and feeder wire which has thick insulation to minimise damage during handling.
- Run twisted loop tails in a plastic tube under the kerb to the footpath and joint properly to a twin flexible cable which is then run back under the footpath to the detector enclosure.
- Use only good quality cable jointing kits to ensure good continuity of conductors and good insulation from earth.
- Before commencing slot cutting check that the road surface is stable (reinstated trenches sometimes bed down

leaving loop wires exposed), that any manhole covers in the vicinity are not likely to move as vehicles pass over them, that the loop will not be within 50 mm of steel reinforcing bars.

13. As you will appreciate it is impossible to mention all eventualities in a paper of this length so, if in doubt, contact the detector manufacturer.

14. Detector design. The electronics of a loop detector must respond to the change in inductance caused by the presence of a vehicle but must ignore other changes in the loop impedance which result from environmental instabilities.

15. It was mentioned earlier that the signal to be detected could be as small as a 0.02% change in total inductance, which is none too easy when you consider that even expensive laboratory inductance bridges rarely have an accuracy of better than 0.1%. In addition the loop detector must operate over a temperature range of -40^0C to $+70^0$C and be capable of rejecting a high level of electro-magnetic interference.

16. There are several ways of achieving this sensitivity, some of which will now be discussed:

- The textbook inductance bridge can be further developed to give the extra sensitivity but this approach is non-resonant and thus has poor noise rejection. The bridge is also likely to be sensitive to changes in the A.C. resistance of the loop brought about by the presence of a vehicle and/or the temperature of the loop and this can cause the detector to 'lock-up'. That is, detect a non-existent vehicle.

- The loop could be used as the inductive element of an inductance/capacitance resonant circuit in a free running oscillator. This overcomes the problem of noise immunity and vehicles can be successfully detected by monitoring the amplitude of the oscillator signal. However, problems can occur if the loop is installed over steel reinforced concrete since some vehicles can then cause the oscillator amplitude to increase instead of the usual decrease, thus causing a 'lock-up'.

- The loop could be used in a free running oscillator as before and the oscillator frequency monitored since this will be inversely proportional to the square root of the inductance. This is perhaps the most obvious approach and is used in many loop detectors worldwide. It does however have one problem. If two adjacent loops have some mutual coupling, the change in frequency caused by a vehicle passing over one loop could cause a detection in the other loop detector if their operating frequencies are not kept far apart.

- The loop could be resonated with a capacitor and then driven, via a coupling resistance, at a fixed frequency, which is close to its resonant frequency. Vehicles can then be detected by observing the phase relationship between the driving voltage and the loop voltage. Although this technique has good noise immunity, does not lock up and runs at a fixed frequency, it does have one disadvantage. The change in phase which occurs is very dependent on the 'Q' or goodness factor of the loop. Therefore a high quality loop will be more sensitive than a poor quality loop.

- The patented Redland system uses two oscillators, one loop oscillator as before and one reference oscillator which is substantially identical to the loop oscillator, but uses a manually tuned inductance in place of the loop. The two oscillators are then loosely coupled by means of a resistor and the phase difference between them monitored. This approach has the advantages of good noise immunity, no lock up, independance of loop 'Q' and, due to the restraining influence of the reference oscillator, only a small charge in frequency when a vehicle enters the loop. A further benefit is the automatic compensation of any temperature effects on the electronics since both loop and reference oscillators are effected in the same way.

17. It is possible to make loop detectors which are manually tuned or automatically tuned. Where the detector is being used for traffic control we believe that manual tuning is better since the unit is given a fixed reference and if it sees an inductance change of more than a certain amount it can "fail safe" and give a permanent detect output to indicate that the loop must have been damaged. In some other applications however, the fail safe facility has no value and simplicity of operation is more important, indicating the need for automatic tuning.

18. All loop detectors should, of course, be capable of tracking environmental drift which can be the equivalent of inductance changes up to 4%. In order to ingnore these large environmental changes whilst detecting small vehicle

signals the detectors must be rate conscious so that they will only react to the comparatively fast vehicle changes.

19. There are fundamentally two approaches to signal monitoring and environmental tracking; digital and analogue.

20. The digital technique is perhaps the most straightforward and is adequate for simple applications. It is combined with a free running loop oscillator whose frequency or period is monitored with counting circuits. The limitation of the free running oscillator has already been described. A further problem with digital detectors is their need to count over many loop cycles in order to achieve the necessary sensitivity, making them very slow and unsuitable for more sophisticated functions.

21. For the time being an analogue detector seems to give the best all round performance but no doubt current advances in the electronic art will soon enable good digital detectors to be achieved, probably using the ubiquitous microprocessor as the controlling element.

22. The future. The Transport and Road Research Laboratory has successfully demonstrated the feasibility of automatic vehicle classification using axle and loop detectors. One of the parameters examined in the classification algorithm is the strength of signal given by the loop detector. This signal strength varies as the vehicle passes over the loop often giving several peaks and troughs. Recent experiments at home and overseas have indicated that by examining the relative amplitudes and positions of these peaks and troughs it may be possible to classify vehicles without the need for axle detectors, thus further improving system reliability.

23. So it can be seen that the loop detector, as well as being the detector of the last quarter century is likely to contribute even more to traffic engineering and surveillance in the next decade. Thanks largely to its high reliability, versatility and accuracy.

REFERENCES

1. D.Tp. Specification MCE 0128

7

Road sensors for traffic data collection

R. C. MOORE, Highway Traffic Division,
Transport and Road Research Laboratory

SYNOPSIS: There is increasing demand for data on the number and type of vehicles traversing a given section of road, and on axle weight, speeds and spacings. Such information is necessary for the development of traffic policies, for studies of road safety and for the planning of programmes of maintenance and new road construction. The paper describes the road sensors employed by the TRRL for traffic counting at the 50-point census plus 21 motorway points, and the 15 point axle weighing programme. Development of an axle counting sensor, suitable for a vehicle count and classification census is described.

INTRODUCTION

Traffic Data Collected by TRRL

1. Vehicle Flow. The first systematic and regular traffic census where the aim was to provide estimates of total travel, was the 50-POINT TRAFFIC CENSUS which was started in 1956. The 50 sites were allocated over the network of trunk and classified roads but excluded unclassified roads. For the first ten years, the data collected were of two types; total daily flows at each site were recorded, and a manual classified count for 3 days each month (16 hours/day) provided the basis for converting the automatic daily counts of total volume, to classified estimates by type of vehicle. The records were extended in 1968 onwards to provide hourly flows rather than total daily flow.

2. The 50 point census was increased to include motorways, the first 3 motorway points becoming established in 1959. The motorway points have gradually been increased, to 21 points, thus data are currently collected from a total of 71 locations.

Sensors in highway and civil engineering. ICE, London, 1981, 73-83

3. Vehicle Speed. The TRRL has carried out a survey on a sample of sections of motorways at 3-yearly intervals from 1970; and in addition more frequent measurements have been taken on the M3 and M4 motorways. These studies employ the Duncan two-speed (moving car) method (ref. 1). The DTp collected speed data using photographic techniques at 49 rural sites in 1975, 1976 and 1977 (ref. 2).

4. Speed data was collected by TRRL at 6 rural road points in 1974 using a radar speed meter. A more comprehensive study at a further 6 rural road points using radar techniques was carried out in 1978 (ref. 3).

5. Axle load. The TRRL has developed a weighbridge system that records automatically, without interference with the traffic, the magnitude of wheel loads passing over it. These weighbridges have been installed, primarily for the purposes of research, at a number of full-scale road experiments and on three motorways. The weighbridges are not suitably sited or sufficient in number for the purpose of defining the nationally representative damaging effect of commercial vehicles on different categories of road. Details of the weighbridge design and installation procedure may be obtained from the report (ref. 4).

6. Vehicle Classification. TRRL has for some years now been developing, with the help of Industry, equipment for automatically counting and classifying vehicles (ref. 5). This equipment detects vehicles by means of a sensor array in the road surface consisting of one loop (vehicle) detector and two axle detectors in each traffic lane. The classifier uses the vehicle sensor signals to determine the speed, wheelbase and axle configuration (including number of axles), length and chassis height (ie the height of the vehicle underpan above the road surface) of passing vehicles, and classifies each vehicle by reference to a look-up table stored in the processor's memory.

VEHICLE COUNTING SENSORS

7. Pneumatic Tube. At the commencement of the 50 point census, pneumatic tubes were employed to record axle counts, from which vehicle counts are obtained (ref. 6). A pneumatic tube produces a change in air pressure due to vehicle tyre contact, which with the use of a diaphragm switch can produce a contact closure or transistor switch output. When correctly fixed upon the road surface, and on lightly trafficked roads, the pneumatic tube sensor is an accurate and useful axle counter. Their disadvantages are associated

with multilane and heavily trafficked roads. On such roads their operational life is short, replacement is difficult and can be dangerous. Traffic lane discrimination (lane counting) cannot be easily obtained. In addition on roads carrying a high proportion of multi-axle vehicles, the conversion of axle count to vehicle count becomes complicated.

8. Surface Coaxial Cable. Suitable coaxial cables are available which may be stretched across or fixed to the road surface. When a vehicle tyre makes contact, the cable is flexed and electric charge generated by friction (triboelectricity). Using suitable electronic detection circuits, signals from this sensor may be used to count vehicle axles. The disadvantages of this sensor are similar to the pneumatic tube, and it is primarily suitable for temporary applications.

9. Inductive Road Loop. This sensor consists of one or more turns of insulated wire embedded in the road (typical depth 5cms), and connected to an electronic sensing circuit. Inductive loop vehicle detection methods are used extensively for vehicle counting and sensing applications ie, traffic light and urban traffic control systems, vehicle presence at gates and parking spaces.

The loop sensor is energised by the electronic sensing circuit, the frequency of operation selected by a capacitor being placed across the loop connections, forming a tuned parallel circuit with an operational frequency:-

$$fo = \frac{1}{2\pi}\sqrt{\frac{1}{LCT} - \frac{RT^2}{L^2}}$$

where L = inductance of the road loop plus feeder cable

CT = total capacitance of the tuning capacitor, loop self capacitance and feeder

RT = total resistance of the loop and feeder, leakage conductance of insulation and eddy current power loss (when vehicle is present), and dielectric loss of the material between the loop wires.

When a vehicle enters the magnetic field of the loop, the following changes occur:

a) eddy currents are induced into the metal of the vehicle. The magnetic field generated by the eddy currents oppose the main field, and produce a change in the relationship between the current flowing in the loop and the voltage applied. Assuming no ohmic loss (the RT term is in practice very small), the presence of a vehicle produces a reduction in the value of road loop inductance.

b) the presence of iron in the vehicle body increases the magnetic flux density and hence increases the road loop inductance. However provided the road loop sensor is of sufficient size, the eddy current effect is predominant, normally by a factor of 10.

c) a very small increase in capacitance occurs, due to the proximity of the loop wire and the vehicle chassis and body.

d) as energy is coupled into the vehicle, so a decrease in the Q of the tuned circuit occurs.

For vehicle counting applications a guide to the configuration and dimensions of inductive loops has been produced by the Department of Transport (ref.7). Inductive loop vehicle counting error is normally within the range of ± 5 per cent. To obtain improved counting accuracy care must be taken to design the positioning and dimensions of the road loop sensors, so that on a multilane road (with one sensor per lane) vehicles in adjacent lanes are <u>not</u> sensed whilst vehicles in each lane <u>are</u> sensed and <u>correctly</u> counted. The use, together with sequential counting logic, of one more sensor than traffic lanes being monitored, referred to as the 'N + 1' system (ref.8), produces a more accurate count on roads where poor lane discipline by road vehicles occurs.

The common problem of both loop counting methods is to record the passage of all vehicles accurately on multilane carriageways regardless of vehicle lateral positioning across the carriageway. The detector threshold level has to be set low in order to detect motorcycles. Two types of vehicle in particular cause further problems, the bus or coach and the aritculated vehicle. The former cause such large inductance changes that they are frequently sensed by adjacent lane loop sensors, whilst the latter cause small changes to the sensor inductance, which are close to the detector threshold so that the counter is triggered 2 or 3 times if the vehicle does not pass squarely over the sensor. These two vehicles, with the exception of motorcycles and bicycles, form the extremes and if the sensor and detector logic can accommodate these two, then all other vehicle categories should also be accurately counted.

For traffic counting accuracy, the multiple detections caused by the articulated vehicle is a greater source of error than that caused by a bus or coach because:

i. there are many more articulated vehicles on the roads (by a factor of 10 higher for example on the M6 motorway)

ii. when an articulated vehicle passes over an 'N + 1' array and does not remain central in its lane, it may produce a continuous detection on the loop it is mainly covering and a double or treble detection in the adjacent lane. The sequential counting logic would then record this as two or three vehicles. An example of single and multi-counting for three vehicle types, for variations in vehicle proximity to the edge of a loop sensor is given in Fig.1.

Despite these problems the road loop sensor has several operational features which make it suitable for traffic detection applications. The detection zones may be varied by loop layout (ref.9), dimensions and positioning, which allows variation to suit the particular road or carriageway. Vehicle sensing is not affected by ambient light levels, dirt, heavy rain or fog, wind and noise levels. The impact of loop sensors on the environment is minimal, since unlike optical, sonic or radar sensors, they do not require supporting poles or cross-arms to mount the sensor.

10. Microwave and Infra Red Detectors. These devices are not currently used for collecting vehicle count data by TRRL or DTp. Their most prominent use is as input sensors for portable traffic light controllers. They may also be used for traffic counting on low flow roads, where the road surface is too poor (gravel) to support inductive loops or other sensors. On multilane roads these devices suffer counting inaccuracies, due to one vehicle shielding another from the detector energy source.

11. Permanent Tribo-Electric Axle Sensor. Over the past three years TRRL have developed an axle counting sensor suitable for vehicle data collection, which overcomes the problems associated with the pneumatic tube and surface coaxial cable. It takes the form of a screened cable encapsulated into a block of urethane, the block being permanently fixed into a slot cut in the road surface. The axle sensor requirements are outlined in further detail:

i) Recovery time: To accurately count each axle, consideration must be given to the time between the passage of successive axles. The axle separation time is a function of vehicle speed and axle separation distance.

$$\text{Separation time} = \frac{\text{axle separation distance}}{\text{speed}}$$

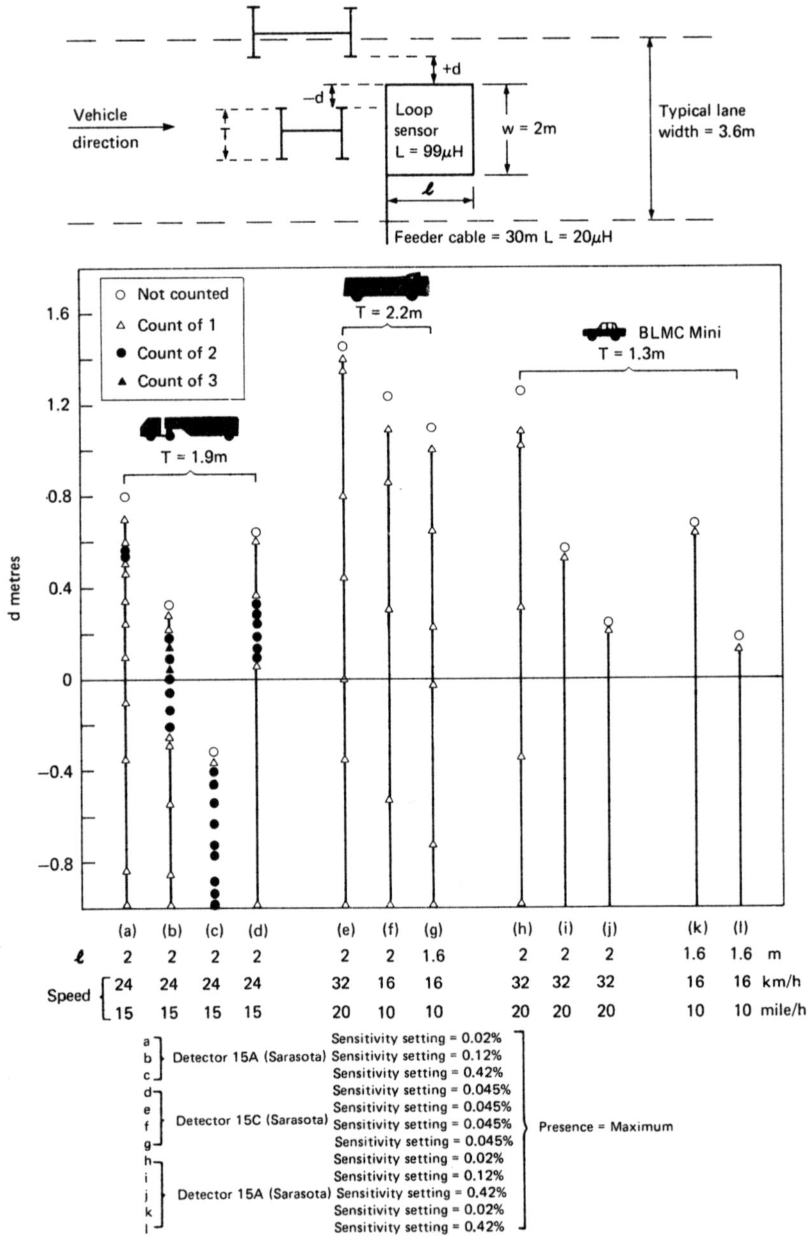

Fig. 1. Vehicle count with variation of vehicle lateral position, with respect to an inductive loop sensor

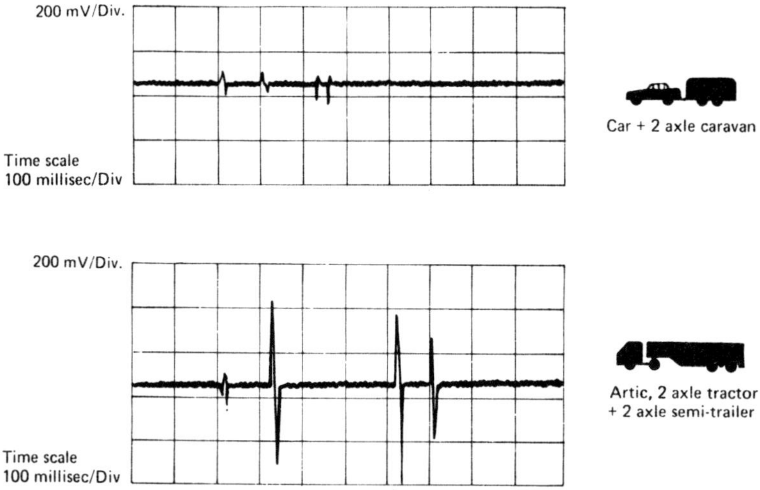

Fig. 2. Example of tribo-electric sensor signals - M4 April 1979

With existing vehicle design, the smallest axle separation distance is found on two-axle trailers or caravans, at approximately 0.75 metre. With such a trailer travelling at 96 kph (60 mph) the axle separation time would be 28 milliseconds and at 113 kph (70 mph) this would reduce to 24 milliseconds.

ii) Response time: The axle detector should be capable of responding within at least one millisecond after being activated and preferably much faster. The requirement to match differing detector response times is smaller.

iii) Sensitivity: The detector must be capable of counting axles over a range of wheel loads (motorcycle up to heavy goods vehicle). In addition it must cater for a range of vehicle speeds 1.6 - 160 kph (1-100 mph).

iv) Vehicle lateral positioning: The detector must be capable of counting axles regardless of vehicle position across a traffic lane, and should have a uniform sensitivity along its length.

v) Stability: Vibration from traffic in adjacent lanes must not cause false axle counting. The detector must accurately function within the temperature range occurring on any road in the United Kingdom (-15 to +55°C).

vi) Installation: The detector must be capable of being installed into road surfaces which are not level, either due to road camber or wear.

vii) Operational life: A minimum operational life of 2 years is required, on a road being repeatedly trafficked by heavy goods vehicles, ie the slow lane of a busy motorway.

Trials have been conducted with a number of differing designs on the M4 and M1 motorways, and other road sites. A total of 53 detectors are currently undergoing life trials at 10 separate road sites.

Early designs had a short life on the slow lane of the M4; however the current designs have lasted over 2 years at the same site without failure. Examples of the signal output obtained from one of these sensors is given in Figure 2.

VEHICLE SPEED DATA

12. All of the sensors previously described, either in pairs or individually (in the case of microwave), may be used for speed measurement. The radar speedmeter is popular and reasonably accurate, ie ± 4 per cent. Inacurracies are caused by multiple reflections, particularly from Heavy Goods Vehicles, and alignment inaccuracies. The portability of the radar unit, plus the fact that no sensors need to be placed in the carriageway, encourages its use. Axle detectors used in pairs, provide an accurate measure of vehicle speed providing the sensors are firmly fixed to the road surface. The permanent tribo-electric axle sensor obviously has an advantage here, and has been found to provide an accurate measure of vehicle speed (±2 per cent). A pair of inductive loops may be used, but they are not as accurate (±3 per cent) as the axle sensors. This is due to vehicles not running squarely over both loop sensors, and sensitivity/detection variations between the loops and detector electronics.

VEHICLE WEIGHT DATA

13. Dynamic Weighbridge. The TRRL design consists of 3 separate weighing platforms, assembled side by side in a frame which is constructed in the roadway. The platforms are 0.6m x 0.6m, and are arranged so that the nearside wheel of heavy goods traffic passes over the centre module. The recorded dynamic load is then doubled to represent the vehicle axle load.

Each weighbridge module consists of two parts, an upper rigid platform of box section and a lower platten, between which are located four strain gauge load cells, one at each corner. The lower platten is secured to the frame by studs, and the upper platform is held to the platten by four spring loaded bolts.

It is important to select carefully a site for a dynamic weighbridge installation. The road should be flat, with a smooth profile and well away from road junctions and curvatures.

VEHICLE CLASSIFICATION DATA

14. To meet the requirement for increased samples of classified counts, associated with speeds, headways and axle weights, TRRL has developed a microprocessor-based vehicle classifier and count unit. A typical sensor array is an inductive loop, and two permanent tribo-electric axle sensors. If axle-weight and total vehicle weight are required, then an additional road sensor (dynamic weighbridge) and electronic decoder unit are needed. The symmetry of the sensor arrangement will allow the equipment to accept two-directional flow, a facility that is important on two-way roads where overtaking is possible and also on unidirectional lanes if for any reason the direction of flow is reversed.

The microprocessor vehicle classifier and count equipment output may be vehicle-by-vehicle or in summary form. The detailed output is essential during calibration and validation of the classification techniques. A summarised output can be obtained at predetermined intervals of time and its content can be specified by means of the software. Additional details on this form of data collection may be obtained from References 5 and 10.

FUTURE

15. There will be a continuing and perhaps increasing need to obtain traffic data. The advent of low cost and vehicle electronic intelligence at the roadside (the microcomputer), opens up the application of data beyond traffic census requirements. The most important merit of the microcomputer is its ability to control processes in real time, and it is this property which will lead to the greatest changes in the use of traffic data. The microcomputer may be used to assess or predict undesirable traffic conditions eg incident detection, high traffic speed in fog or ice conditions, undesirable concentration of axle loads approaching a bridge undergoing repairs, and call up the appropriate control measures.

The need to improve road sensors to provide reliable and accurate data for the microcomputer will continue. The advent of the piezo-electric load sensor, which despite some operational problems, offers the strong possibility of obtaining a reasonably accurate measure of axle weight at relatively low cost.

ACKNOWLEDGEMENTS

16. The work described in this paper forms part of the programme of the Transport and Road Research Laboratory and the paper is published by permission of the Director.

REFERENCES

1. DUNCAN N C, S L SUMNER and J S SHREWSBURY. Measurement of the speeds of cars on motorways in 1976. TRRL SR 326.
2. Road Accidents in Great Britain 1977. HMSO 1978.
3. NICHOLL J P. Factors associated with the distribution of vehicle speeds at points on single-carriageway rural A roads. Transport Studies Group, University College, London. July 1980.
4. TROTT J J, and J W GRAINGER. Design of a dynamic weigh-bridge for recording vehicle wheel loads. TRRL LR 219.
5. MOORE R C. An automatic method to count and classify road vehicles. OECD Seminar, TRRL 1980.
6. BLACKMORE D H. Operation and maintenance of the Fischer and Porter punched-tape counter. RRL LR9.

7. GERAGHTY F. A guide to the configuration, installation and termination of inductive loops for traffic counting purposes. Technical Guide MCH 1213, TCC.

8. GATER P M, and R K DULEY. Current automatic traffic counting practice. ICE Traffic Data Collection, April 1978 - Paper 6.

9. DAVIS L W, K G DEAN, D E MACDONALD, and D J MORRIS. Vehicle detector loop configurations. AWA Research Laboratory (NSW Australia), Report No R76-68.

10. HILLIER J A, MATHEWS D H and R C MOORE. Trends in traffic data collection and analysis. ICE Traffic Data Collection, April 1978 - Paper 9.

8

Detection of overweight vehicles at Clifton Suspension Bridge

B. J. THORPE, BSc(Eng), MICE, Director, Howard Humphreys & Partners

SYNOPSIS. Installation of permanent weighbeams in the approaches to the Clifton Suspension Bridge has ensured detection of overweight vehicles before they reach the bridge itself. The signal generated by an overweight vehicle provides a visual warning for the driver, an audible alarm for the bridge staff and also stops the bridge toll barriers from opening, thus preventing the vehicle from reaching the critical suspended deck section.

INTRODUCTION

1. The Clifton Suspension Bridge is well known to both engineers and the general public for its association with I. K. Brunel, and the general layout is shown at Fig. 1. Apart from being important historically the bridge still performs a valuable service today by providing a crossing of the Avon Gorge linking a number of residential areas with Bristol.

2. The bridge is a major tourist attraction with more than 3 million vehicle crossings a year and maximum flows of a little over 1,000 vehicles per hour in one direction at peak periods despite the narrow 5m carriageway.

3. The history of the bridge is well recorded (ref. 1,2,3) and at the time of its opening in 1864 it was intended to carry the typical traffic of the day: mainly horses and carts, coaches and farm animals. For this purpose a wooden deck of 125mm (5 inches) deep timbers running longitudinally over the wrought iron cross girders sufficed. Transverse planks 50mm (2 inches) thick were used as a wearing course which could be renewed when they had been worn through by the iron-rimmed wheels. The Clifton Suspension Bridge Act, 1861 provided for a limit of 28 tons

Sensors in highway and civil engineering. ICE, London, 1981, 84-94

for the total live load on the bridge and also stipulated that the gross weight of individual vehicles was not to exceed 6 tons.

4. Though the deck timbers have been renewed at intervals the original construction has been retained but with the addition of mastic asphalt surfacing over the transverse planks to waterproof the deck and form a new wearing course. Calculations have shown that critical sections in the cross-girders would be overstressed by modern vehicles of 6 tons gross weight and Byelaws now in force under the Clifton Suspension Bridge Act, 1952 provide instead for a maximum axle load of 2.5 tons.

5. While the bridge staff have long had the power to require the driver of a vehicle they believe to be overweight to produce a weighbridge ticket to prove compliance with the Byelaws - which in effect has meant that such vehicles have sought an alternative crossing of the river - protection of the bridge has, until recently, relied on the high standard of the advance warning signs and on the ability of toll staff to identify overweight vehicles visually.

6. Until 1975 all tolls were collected by hand, by staff stationed in the approaches to the bridge. During the day these staff collected tolls on both approach roads thus being able to check for overweight vehicles before they reached the suspended deck section. At night, for security reasons and because flows reduced to very low levels, tolls were collected only at the Clifton end of the bridge from vehicles travelling in both directions. This left the possibility of an overweight vehicle being identified only after it had crossed the bridge from the Leighwoods direction.

AUTOMATION OF TOLLS

7. In 1975 automatic toll collection was installed (see Fig. 2) which largely eliminated the need for bridge staff to work outdoors and improved the financial control over toll collection. Because staff were no longer stationed in the carriageway it also increased the chance of an overweight vehicle passing undetected.

8. While suitable weighbeams were not available at that time provision was still made in the toll equipment for future utilisation of a weighbeam signal so that overweight vehicles would cause the barriers to remain in the closed

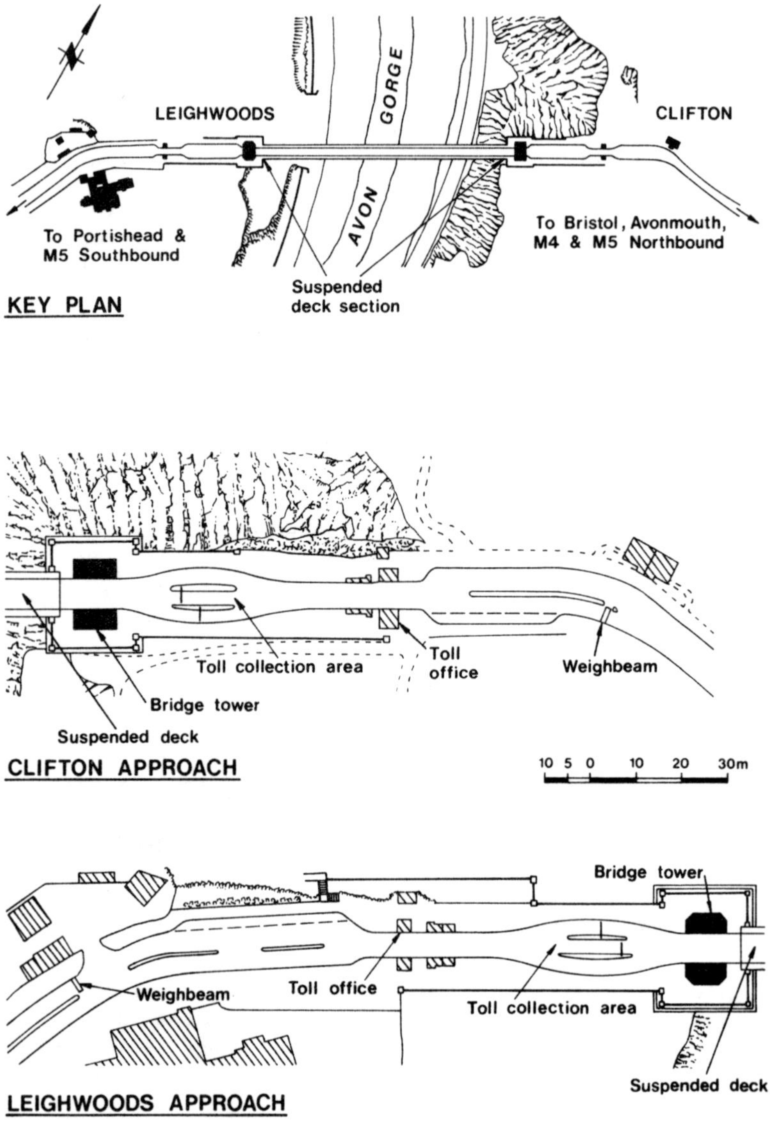

Fig.1. Clifton Suspension Bridge and approaches

position even when the correct toll had been paid.

WEIGHBEAMS

9. By 1977 it became clear that weighbeams developed by Weighwrite Ltd. to meet the needs of the Department of Transport offered a sufficiently reliable product to justify installation on a permanent basis at the bridge.

Siting

10. One of the most difficult problems was the siting of the weighbeams. To ensure that overweight vehicles were intercepted before they reach the suspended deck section the weighbeams had to be located in the approach roads, one at each end of the bridge.

11. By placing them in the dual carriageway sections of these roads (see Fig. 3) it was possible to be certain that both wheels were weighed together on each axle of every vehicle and also prevent false alarms through simultaneous loading from the wheel of a vehicle passing in the other direction.

12. While the manufacturer's requirements for a normal installation included a straight and level site, the existing highway alignment could not be more than slightly amended with the result that the Clifton weighbeam is installed with a crossfall of 5% and on a horizontal radius of 30m, while at Leighwoods the longitudinal gradient is 2%.

13. Calculations indicated that these undesirable geometrical features would not affect the perceived loads by more than a few percent - probably to a lesser extent than the dynamic effects mentioned in paragraph 33.

14. As braking forces were known to be capable of causing substantial changes in axle loads, trial runs were made and vehicles observed to determine zones where braking was least likely to occur.

15. The final sites chosen (see Fig. 1) are the inevitable compromise that has to be accepted in civil engineering: far enough away from the bridge to be outside the zones of heavy braking in the approaches to the toll areas, far enough away to allow the secret signs to give adequate warning to stop an overweight vehicle but in locations where geometric effects are avoided as far as possible.

Fig.2. Automatic toll collection at Clifton end of bridge

Fig.3. Weighbeam in Leighwoods approach road

Standard weighbeam

16. The standard product consists of a 3.073 x 1.042m rectangular fabricated steel weighing platform supported at each corner by hydraulic load cells and set flush with the road surface in a shallow concrete pit.

17. Both the beam and the road surface for some 3m either side must be level to close tolerances and vehicle speeds during weighing must not exceed 3.5 km/h.

18. Signals from strain gauge type transducers in the load cells are normally analysed to provide digital display of individual axle loads which are also printed on paper tape to give a permanent record together with gross vehicle weight and time and place of weighing.

19. In this configuration the equipment is capable of giving accurate axle loads (within ±50 kg up to 15 tonnes) suitable for law enforcement purposes and for use by industry in checking vehicle loadings.

Modifications

20. A vehicle speed survey showed that cars travelled at up to 50 km/h and lorries at up to 35 km/h at the location selected for the Leighwoods weighbeam compared with the 3.5 km/h maximum specified by the manufacturer.

21. Because the object was to avoid interference with traffic flow, modifications were necessary to the analogue circuitry to allow production of a satisfactory signal for these vehicle speeds although this was accompanied by some loss of accuracy. This loss was expected to be of the same order as the few percent anticipated for the geometrical effects described in paragraph 13.

22. A special control console was designed to meet the particular needs of the bridge staff and incorporated the following:
- push-button switch for mains power supply: illuminated when power on,
- push-button switch to cancel an alarm signal: illuminated under alarm condition,
- calibrated meter for axle loads up to 5 tonnes,
- four push buttons to set zero on the meter and generate standard signals for checking functioning of equipment,
- key controlled inhibit switch to prevent alarm circuits operating during certain maintenance and testing operations.

23. To improve their life the weighbeams were galvanised and calcined bauxite anti-skid treatment was applied to their upper surfaces. Because of the longitudinal and transverse gradients at the weighbeam sites special chocks were used to prevent movement under traffic conditions.

Functions

24. Following modification the weighbeams now weigh every axle of every vehicle at normal traffic speeds. The signals are analysed in the control consoles which are located next to the toll barrier consoles in the old toll houses.

25. When an overweight vehicle is sensed relays are closed which

- operate an alarm bell to alert the bridge staff,
- operate a secret 'STOP' sign at the toll house to halt the driver before he reaches the toll collection area,
- lock the toll barriers in the closed position immediately, if they are already down, or at the end of their cycle if a vehicle is passing through.

Thus, even a vehicle which has failed to see or has ignored the 'STOP' sign at the toll house is still prevented from crossing the bridge.

26. The bridge staff can cancel the alarms and then release the barriers in each of the two toll lanes independently, so cars which are being held up by the presence of an overweight vehicle can be cleared through one lane while the overweight vehicle is held in the other.

27. The test facilities allow the proper functioning of the weighbeams to be checked at the start of each shift.

28. The equipment is virtually 'fail-safe' as a power failure automatically causes an alarm condition and other failures, such as in a load cell, generally lead to the same result.

Installation and costs

29. Because of the importance to daily traffic of the route over the suspension bridge, the weighbeams were installed at one end of the bridge at a time with one-way controlled traffic working. At night short Bailey Bridge sections were erected to allow normal two-way traffic. In addition, the contractor's working hours in the carriageway were restricted to the period between 9.30 am and 4.00 pm to avoid peak traffic periods. While these restrictions

severely limited the contractor's working time, they had the desired effect in that delays generally to traffic were at a minimum while the main peak period traffic was virtually unaffected.

30. The measures taken to avoid disruption to traffic and the apalling weather during the first four months of 1978 when the work was carried out were reflected in the civil engineering costs under what was essentially a daywork contract- these costs including all necessary re-surfacing, ductwork and drainage connections amounting to £15,852. Purchase of weighbeams and commissioning by the manufacturer cost £13,014 which included the specially designed control console and all modifications to the standard weighbeams.

Calibration and site tests

31. Apart from initial vehicle speed and braking surveys a further series of tests was carried out on site in April 1978 as part of the commissioning procedure to ensure satisfactory performance of the weighbeams following calibration by the dead weight method in the factory.

32. A lorry was selected which had its heaviest, front axle very close to the 2.5 ton limit. After checking the actual weight for the particular fuel load, driver weight etc. on the public weighbridge a static check was made at each weighbeam to ensure the perceived axle load was the same as the public weighbridge figure.

33. The lorry was then driven over the weighbeams at a variety of speeds to well beyond those normally used by lorries in the approaches while the weighbeam signals were monitored by high speed ultra-violet recording oscillograph. The results (see Figs. 4 and 5) were very satisfactory and, as expected, the geometric configuration and dynamic effects resulting from irregularities in the carriageway profile caused both the maximum perceived load and the scatter to increase with speed.

34. At normal lorry speeds in the approaches of around 15-20 km/h at Clifton and 30-35 km/h at Leighwoods the perceived load was about 2.8 tonnes. To allow a margin for braking effects and the varying dynamic responses of other vehicles the alarm level was set at 3.0 tonnes.

35. Finally, a Clifton College coach which is the heaviest vehicle known to use the bridge regularly, with a rear

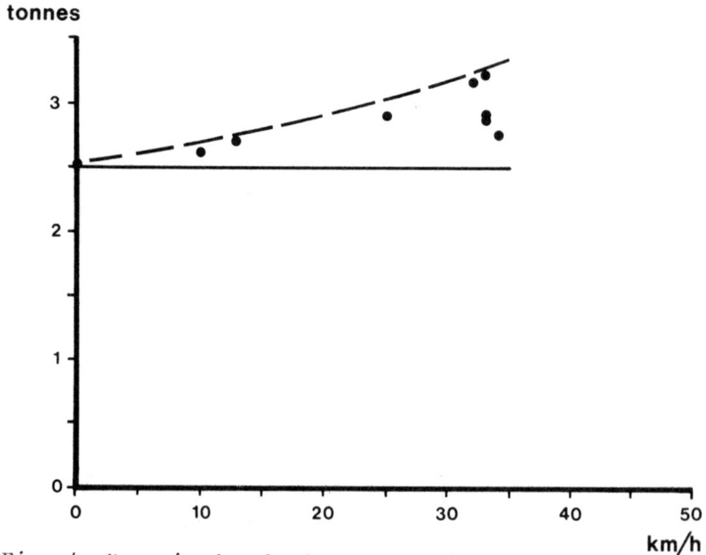

Fig. 4. Perceived axle loads at Clifton

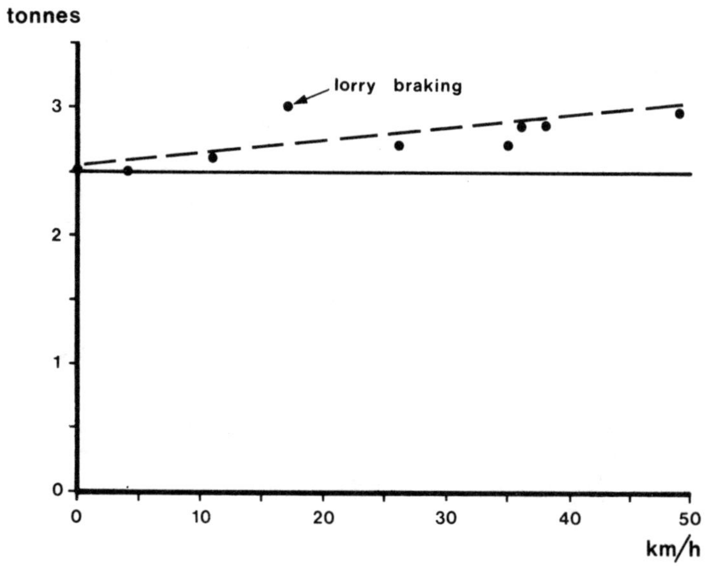

Fig. 5. Perceived axle loads at Leighwoods

axle weight of 2.35 tonnes unladen, was driven over the weighbeams at various speeds not exceeding 20 km/h and was able to pass consistently without setting off the alarms.

Results in use

36. In the first three full months after the weighbeams came into use 64 overweight vehicles were detected and turned away from the bridge.

37. This rate of well over 20 vehicles per month fell rapidly, presumably as knowledge of the presence of the weighbeams spread, and over the next 6 months an average of 7 vehicles was turned away each month.

38. In the same period a year later the number of overweight vehicles was down to an average of 5 per month: a level which now appears to be steady.

39. Of the 224 overweight vehicles detected between April 1978 and the end of August 1980 the majority were lorries (including articulated, refrigerated and container lorries) and tourist coaches from many parts of Europe.

40. Because the residential area at the Leighwoods end of the bridge is itself subject to a 4 ton (gross) weight limit (except for access) the number of overweight vehicles arriving at that end of the bridge is only 60% of those arriving at the Clifton end.

Maintenance

41. Through a contract arranged with the manufacturer the weighbeams are lifted out annually for inspection and cleaning and, after replacement, their calibration is checked using a vehicle with a known axle weight.

42. The contract provides for emergency attendance for repairs and a stock of selected spare parts is maintained at the bridge, such as matched transducers, to avoid delivery problems.

43. The equipment has functioned continuously since installation with only one significant problem which has been the failure of transducers on a number of occasions through lightning strikes to the bridge. It is hoped that improved earth bonding recently completed will minimise this problem in future.

CONCLUSIONS

44. The weighbeams have shown a high level of reliability in continuous use after well over two years of operation during which time nearly 8 million vehicles have been weighed, without inconvenience to the drivers.

45. Despite good advance warning of the weight limit, 224 of these vehicles were found to be overweight and turned away.

46. The significant reduction in the number of vehicles intercepted after the first few weeks suggests a 'learning curve' and this points strongly to there having been a number of vehicles (probably of local origin) which had regularly used the bridge despite being overweight.

47. By preventing overweight vehicles crossing the bridge cumulative damage to the structure and surfacing has been avoided with a corresponding reduction in maintenance costs.

48. Probably the most important feature of the installation is the knowledge that day and night protection against the passage of any grossly overweight vehicle is assured: such a vehicle being capable of causing serious damage to the deck over the whole length of the bridge which would require heavy expenditure to rectify as well as prolonged closure of an important local traffic route.

ACKNOWLEDGEMENTS

49. The author wishes to thank Weighwrite Ltd. for their assistance and the Clifton Suspension Bridge Trust for permission to publish this paper.

REFERENCES

1. BARLOW W. H. Description of the Clifton Suspension Bridge. Minutes of Proceedings of the Institution of Civil Engineers, 1867, Vol XXVI, February, 243-257.
2. PORTER GOFF R.F.D. Brunel and the design of the Clifton Suspension Bridge. Proceedings of the Institution of Civil Engineers, Part 1, 1974, 56, August, 303-321.
3. PUGSLEY Sir Alfred. The works of Isambard Kingdom Brunel. Institution of Civil Engineers, London, 1976, 51.

Discussion on Papers 4–8

Mr F.J. GERAGHTY (*Paper 4*): The advent of micro-computers offers traffic engineers the versatility to devolve intelligence down to the transducers, the signal conditioning the processing and finally the display and output stages of control and monitoring systems. It enables the benefits of programmability to be built into systems. In the past, many systems have been produced with limited flexibility. They have been built with dedicated hardware which required the technician to get in with the soldering iron to carry out modifications. Although he has successfully carried out the modifications there have been side-effects which were not intended - either dry solder joints or wires in the wrong place - and always the problem then of re-commissioning the system.

The advent of computers themselves improved the situation. It enabled transducers to be input via transmission networks to the central processor and, with fairly sophisticated programs, to extract from these transducers information relating to the behaviour of the system. However, there were some disbenefits associated with this approach. It usually required highly skilled staff to maintain and operate these systems. Changes in the system required major changes to the software, and it is in software - whether in mainframe computers, mini-computers or even micro-computers - that costs tend to escalate. Although a major portion of overall cost occurs in the capital expenditure and design of the system an area often ignored is the maintenance cost (hardware and software), which continues for the life of the system. In centralized computer systems the transmission networks and the software require a great deal of skill, attention and money to keep them fully operational.

However, the advent of micro-computers does offer traffic

engineers the chance to distribute intelligence and to minimize the complexity of tasks which each individual element is required to undertake. It also offers the chance of putting together a universal hardware - the kind of system that can then be sub-equipped or extended on a modular basis to undertake functions under software control. Such operating software systems are called firmware because they are located in non-destructive memory modules. The data or applications software are located in random access memories and are protected against power failure.

However, the configuration of hardware and software is only part of the problem. It is easy to enter the field of deferred design - buy the equipment first and work out the system afterwards. Such an approach causes problems: it sometimes excuses the system designers from properly resolving the system and problems.

There are two principal users of data: the system and the operators. The first questions to be determined are what data are required for the system and how are they to be used? The system designer is presented with a top-down procedure: he must design the inputs and outputs from the primary source to the end user, and then break them down into modules so that each task is related to all the other tasks in the system. With programmable systems there is a tendency to collect, process and display far more data than are necessary for the purposes of the system or the user. Because data can now be collected readily it is felt that the data are needed. 'What data do your need for the system?' is sometimes answered by the user with 'What data can I have?' This is a closed-loop situation. The questions to ask are: 'Are your data-collection data necessary? How can you minimize them? How can you reduce them? What are the benefits and disbenefits of not having or having these data?', and most important, 'How is the system going to operate in a failure mode?' It is very difficult, with sophisticated complex systems to define all the failure modes. Consider how the system will take off when a parameter is given either as totally failed or distorted. These considerations are as important as the actual hardware and software used in the system. Unless the system is completely understood there is little point in buying sophisticated hardware because sophisticated problems will result.

An old systems engineering adage advises always keep the aims and objects of your system in view. It is very easy to forget when involved in overcoming the problems - to the extent that problems are created for the sport of solving them.

Finally, the role of the operator in the system is very important. It is dangerous to present him with unnecessary data because if they are displayed, he will do something with them and when they are missing or distorted he will take some action. There is the further problem that if the operator is using such data then his attention is distracted from other areas.

With micro-computer systems nothing more than a VDU and hard-copy logger are required. These can be very sophisticated, but it is essential to decide what is necessary to display on them.

Mr G. FUDGER (*Paper 5*): There are many problems involved in the selection and design of road tunnel environmental systems and their controls and thus their sensors. The conditions inside a road tunnel are very aggressive; being damp, dirty, highly corrosive with turbulence from passing vehicles creating the conditions for premature failure of any poorly designed equipment made from inadequate materials. Because of the difficulties of access, the reliability and maintainability of equipment and the stability and accuracy of sensors should be emphasized. The latter qualities have become more critical since the rise in electrical power costs in recent times has increased the penalty for inefficient operation. Tunnel managers are seeking economies as is being demonstrated, for example, at the Tyne Tunnel where a new carbon monoxide sensor, highly robust and stable, has enabled much closer control of the ventilation. The saving in electricity should pay for the cost of the instrument in about 2½ years. With the incentive to reduce costs and maintain them at a minimum consistent with safety in the tunnel, sensors will continue to be critically examined and improvements sought.

It has been said that in order to have computer-based controls the user must specify the exact requirements to the system designer. One difficulty is that, because of their individual nature, the operator has to gain experience with his tunnel in order to know which problems are important, and these might change with time. Thus he and the tunnel designer may not know what is needed in detail before the tunnel is opened. This is illustrated in Fig. 1 which shows the decrease in power consumption against the increase in traffic at a major tunnel over the years (ref. 1).

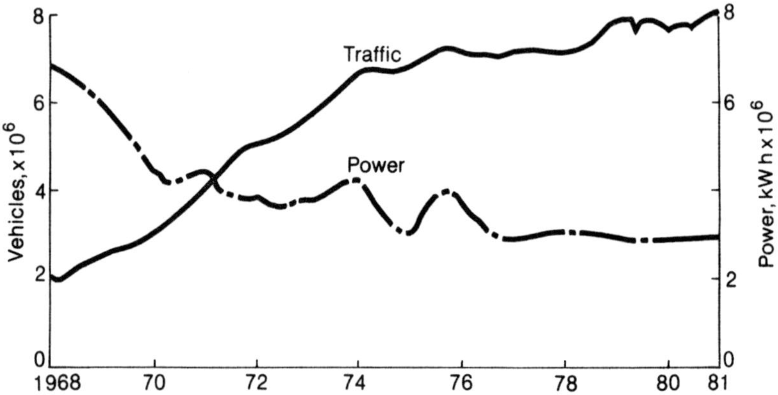

Fig. 1. Tyne vehicular tunnel, traffic and power consumption: moving annual totals

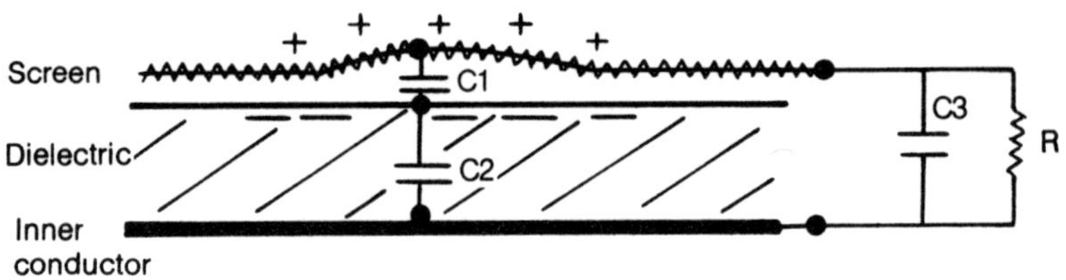

Q = charge between screen and dielectric

Q_1 = final charge on C_1

Q_2 = final charge on C_2

$$\frac{Q_1}{C_1} = \frac{Q_2}{C_2} \text{ and } Q = Q_1 + Q_2$$

$$Q_2 = \frac{Q \cdot C_2}{C_1 + C_2}$$

$$\text{Maximum voltage across } C_3 = \frac{Q_2}{C_3}$$

$$\text{Generated voltage} = \frac{Q}{C_3} \times \frac{C_2}{C_1 + C_2}$$

Fig. 2. Screened cable - voltage generation

Mr R.C. MOORE (*Paper 7*): The tribo-electric axle sensor has been developed to provide an accurate count of vehicle axles at permanent traffic-census sites. Tribo-electricity is defined as frictional electricity, a separation of electric charge which results from the rubbing together of different materials. In the world of electrical cables, tribo-electric noise may be defined as that component of electrical noise generated in a cable, by the mechanical stressing of the cable.

At TRRL we have designed a sensor to record the passage of vehicle axles, which utilizes the tribo-electric effect. Our knowledge of the precise mechanism by which voltage is generated by cable stress is limited, but Fig. 2 indicates the relationship as we understand it.

The sensor is manufactured by encapsulating a suitable screened cable into a block of urethane. The block is subsequently epoxied into a shallow slot cut into the road surface. The voltage amplitude produced when a vehicle tyre contacts the sensor is dependent on the cable type, cable depth within the block, shore hardness of urethane, and the excitation force. One of the major aims of this work has been to produce a low price sensor with an operational life in excess of two years on a busy motorway or trunk road. Over 50 of these sensors are undergoing life trials and Table 1 illustrates the operational life so far obtained at one site.

Table 1. Axle sensor trial - A33 Winchester (1/4/81)

Construction	Traffic lane/ position	Average weekly flow, % commercial	Installed	Failure
S2/I80/D85	1A	76 500 (31%)	14/10/79	None
S2/I80/D58	1B	76 500 (31%)	14/10/79	None
S2/I80/D83	2A	61 000 (15%)	14/10/79	None
S2/I80/D84	2B	61 000 (15%)	14/10/79	None
S2/I80/D81	3A	60 500 (22%)	7/10/79	None
S2/I80/D82	3B	60 500 (22%)	7/10/79	None
S2/I80/D78	4A	78 000 (27%)	7/10/79	20/3/80*
S2/I80/D79	4B	78 000 (27%)	7/10/79	None
S2/T80/D108	4A	78 000 (27%)	20/ 3/80	5/12/80†
S2/T80/D131	4A	78 000 (27%)	11/ 1/81	None

* Failure due to water ingress - poor cable end seal
† Failure due to split urethane block

Mr J.W.M. VALLIS (*County Engineer, Warwickshire County Council*): Although there is less major construction on roads, maintenance has become of paramount importance. The development of loop detectors has meant that traffic signals in town centres are more efficient, and signals can now be located in rural areas where they would not have been put before because of the accident risk. They will also be used for traffic prediction where reassessment is required, although further temporary means of assessing traffic flows need to be developed.

The development of the tribo-electric axle sensor into a more simple portable type which could also pick up traffic classification, would be of great assistance to highway engineers.

The facility to weigh axles on motorways (particularly those of heavy traffic) is greatly needed for the planning of motorway reconstruction and for tests to assess how they stand up to predicted loads. There are weighbridges in position (there are some on the M1) but the stage is being reached where weighbridges are a bit heavy for the job.

One of the problems with the loop detector, particularly in towns, is the damage caused to them by roadworks.

The technology to produce sensors which can flash signs up is needed to deal with the problem of vehicles hitting low bridges: an aqueduct in Warwickshire was hit four times during November and December 1980. Sensors have been devised to go on the rear of dump trucks (ref. 2). If the trucks reverse and they come to a hole a message ensures the brakes are applied automatically. The same principles can be applied to safeguard low bridges.

Mr GERAGHTY: Trials are being conducted by the Department of Transport under the auspices of the Bridges Engineering Division, to look at the whole character of 'bridge bashing' It is particularly related to bridges carrying high speed trains. Currently, the Traffic Control and Communications (TCC) Division of the Department of Transport are conducting trials of the equipment systems to detect the overheight vehicles, divert them and also monitor whether they proceed beyond the danger area and strike the bridge.

There are now equipment systems available using laser methods which can detect the presence of high vehicles, note their direction of travel, monitor their speed, light the sign for them and extinguish it only when the vehicle has passed. In Wiltshire on the approach roads to Salisbury there is a system which alarms on-coming vehicles that a high vehicle is in the middle of the road and warns them that they should allow this vehicle to progress. There is also a sign which warns a vehicle to turn back. Where diversion routes exist the vehicle is warned sufficiently far in advance so that it can pick up the diversion route and go along it.

On bridges lasers are used for the communication from the overheight detector to the bridge sign; this eliminates the need to lay cables to connect the detector to the sign. These systems have operated successfully over a distance of 300-400 m. In Bristol trials have been conducted from our headquarters to the regional office from a line of site from the top of one building to the top of another, over a distance of more than a quarter of a mile, in signals communication. These systems can be used to convey information and messages and reduce the need to lay cables on carriageways, but there is not a marketable system available yet.

Mr M.J.H. CHANDLER (*Head of Urban Traffic Control & Signals, Greater London Council*): I would like to mention the problem of loop detector performance at vehicle-actuated traffic signals. Probably around one-half of an authority's installations have some fault present in the detector units or loop cables at any given time. The problem is not so much in the concept, but in the fact that the technical standards are not high enough.

A configuration needs to be introduced that has a better tolerance to environmental factors on site, and which reduces the risk of damage and hence maintenance requirements. Examples of this are self-tuning detectors, armoured feeder cable, better joints, better insulated loop wire, and a single loop for each approach.

Alternatively, with microprocessor controllers, multiplan fixed-time operation with no detectors would be attractive at many sites.

Mr GERAGHTY: The TCC Division of the Department of Transport, are responsible for drawing up setting standards and issuing specifications. The Welsh Office have conducted investigations for their requirements, as has the Scottish Office. However, no matter how good standards specifications are, it is only by supervision that those standards will be enforced and this will require additional expenditure.

As to the elimination of two loops and having one loop instead of three, that is a matter for the Department of Transport to see whether or not there is any danger inherent in doing this. Certainly, the adapted system has tangible benefits from going from fixed-time planning to real-time control.

Mr MOORE: With regard to inductive loop sensor failure, good specifications covering installation technique and materials need to be supported by good workmanship and supervision. It was suggested that a single loop per approach lane at vehicle-actuated traffic signal junctions could improve the reliability of detection. This would certainly be feasible, so that in the event of one loop sensor failing on, say, a three-lane approach road, counting could continue on two out of the three lanes, and a weighting factor applied for the missing lane count. Such a scheme would need further research.

The Federal Highway Administration (FHA) offices of research and development in Washington, USA, have developed a new self-powered magnetometer for vehicle detection. This battery-operated sensor detects a vehicle's magnetic signature and transmits this information to a nearby receiver. The estimated battery lifetime for a vehicle flow of 20 000 vehicles a day is greater than one year. Experience with loop sensors shows that most of the damage, caused by subsequent road construction works, occurs to the feeder cables. Getting rid of the feeder, as the FHA are doing, would be a valuable contribution to improving sensor reliability.

With regard to vehicle axle weighing, TRRL are conducting trials with a French piezo-electric cable sensor.

Mr G. JONES (*Partner, Wootton Jeffreys & Partners*): A sensor which is cheap, has no construction problems, does not wear out and is very portable - and therefore ideal for many applications - is the video camera. Wootton Jeffreys & Partners have developed a system called VISTA which has been used for various applications, linked to a microcomputer, so that an automatic analysis can be quickly obtained from it. A time pulse is fed from the video signal against which various events are recorded. This can be used for classified or directional turning movements, at different sorts of junction, for pedestrian and traffic movements. It has been used for road studies and at show and exhibition sites. It gives a cheap, quick way of getting data out of the system. We are developing this further to obtain various inventory information, such as size information, spacings, physical considerations of the road and so on. All this can be analysed by the computer.

I understand only one source has been found by the Department of Transport to check the accuracy of various counting methods, including loops, classified counts, manual, and so on, in their analysis of dealing with uncertainty. This was undertaken by us for the GLC on Blackfriars Bridge, where the VISTA system was used to give complete coverage. VISTA also has the facility to go back for re-check and review when errors occur.

Mr GERAGHTY: In paragraph 16 of Paper 4 reference is made to the use of closed circuit television as part of the integrated system. Line and frame snatching techniques are now being actively investigated by the Department of Transport.

REFERENCES

1. Allenby, H. Operational feedback. 'Proceedings of the 3rd international symposium on aerodynamics and ventilation of vehicle tunnels.' British Hydromechanics Research Association, Cranfield, Discussion paper, 1979.

2. Byles, R. Sensing danger behind large trucks. 'New Civil Engineer', 1981, Jan., no. 424, p. 16

9

Measurement of road surface shape using a laser sensor

P. G. JORDAN, BSc, Principal Scientific Officer, Transport and Road Research Laboratory

SYNOPSIS. A contactless laser sensor has been developed that is capable of measuring vertical height from the road surface up to 3300 times per second. It has a resolution of 0.282 mm over its working range of 72 mm, which is centred at a distance of 270 mm from the sensor. The sensor has been employed in the development of computer controlled systems for measuring, at traffic speed, the longitudinal profile, the macrotexture and the wheel-track rutting of a road surface. The operation of the sensor and its applications to the measurement of surface shape are described.

INTRODUCTION

1. Information on road surface shape is required by the road construction engineer to assess compliance of new construction with specification, and by the maintenance engineer for assessing maintenance needs and priorities. In recent years research effort has been devoted to the development of high-speed profile measuring equipment to provide the required data.

2. The Transport and Road Research Laboratory had developed initially a mechanical wheeled-profilometer that served a useful function for many years. But the vibration, wear and thermal problems that result from the use of mechanical sensors limits their performance for profile measurement.

3. Consequently the Laboratory has developed (ref. 1) a contactless laser sensor to measure displacements from the road surface. The sensor has applications for the measurement, at normal traffic speed, of longitudinal profile,

Sensors in highway and civil engineering. ICE, London, 1981, 105-117

macrotexture and wheel-track rutting.

CONTACTLESS SENSOR

4. Description. The essential features of the laser sensor are shown in Fig. 1. Laser light projection and receiving units are fixed to a rigid beam, such that the projection and receiving axes are orthogonal. The lens in the projection unit focuses the light from the semi-conductor laser onto a small area of approximately 0.3 mm^2 of the road surface. A proportion of this light is diffusely scattered by the surface and a fraction of the scattered light is collected by the lens in the receiving unit and is focused onto a linear array of photo-diodes.

5. When the surface is displaced vertically relative to the sensor, the illuminated area of surface moves along the axis of the projection beam. Because the projection and receiving axes are orthogonal, the image of the illuminated area formed on the diode array moves along the array and remains in focus as shown in Fig. 1. By detecting which diode in the array has maximum light intensity the vertical distance of the sensor from the surface can be calculated using the displacement calibration data for the sensor.

6. The laser source produces infra-red light at pulse rates up to 3300 per second. The sensor has a resolution of 0.282 mm, determined by the spacing of the diodes in the diode array, and has a working range of 72 mm centred at a point 270 mm from the sensor; accuracy of measurement is ±0.282 mm over the full working range.

7. Operational factors. The operation of the sensor depends upon sufficient light being received by the diode array to generate a signal from an individual diode. To avoid spurious signals being generated by electronic noise or strong ambient light, a threshold level is set for the diode below which no signals are generated; in practice, an opaque skirt is also provided to protect the array from surface reflections of strong sunlight.

8. The intensity of the light scattered from the road surface is influenced by the shielding effect of the surface macrotexture and by the reflectivity of the surface. Because road surfacings have a spectrum of texture depths and reflection coefficients, the intensity of the scattered light varies with surface type and condition. On occasions, insufficient light reaches the diodes and a lost measurement or 'dropout' occurs. The proportion of 'dropouts' found on

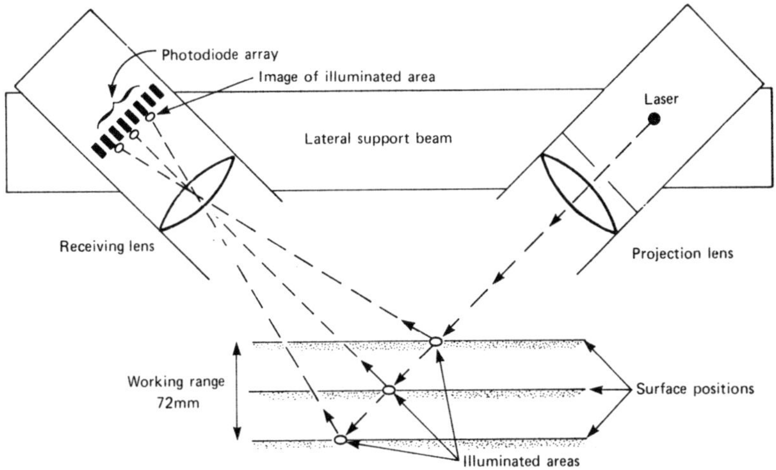

Fig. 1. Essential features of laser sensor

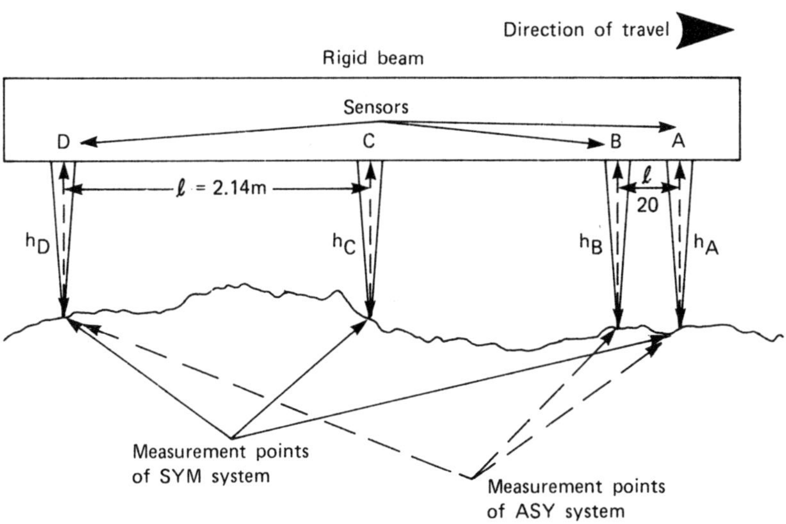

Fig. 2. Sensor configuration on rigid beam of high speed profilometer

different surfacings range from zero on smooth concrete to 50 per cent on a pervious macadam surfacing. Because the sensor's measurement rate is high and because 'dropouts' rarely occur in sequences of longer than two (ref. 1), and are randomly distributed over the road lengths being measured, proportions of 'dropouts' up to 70 per cent are tolerable for profile and texture measurements.

HIGH-SPEED PROFILOMETER

9. Description. The high-speed profilometer designed at the Laboratory (ref. 2) uses four laser sensors fitted to a 4.5 metre-long rigid beam in the configuration shown in Fig. 2. The sensor configuration provides two profile measuring systems, the symmetric (SYM) and the asymmetric (ASY); each measures a different range of wavelengths. Profile features with wavelengths greater than 10 metres are measured by the SYM system (sensors A, C and D in Fig. 2); this computes the average profile height over 2.14-metre lengths every 2.14 metres to give the long-wave profile of the road. The ASY system (sensors A, B and D in Fig. 2) determines the profile in the wavelength range 0.3 to 20 metres in terms of the average profile heights over 0.107-metre lengths every 0.107 metres along the road. A single composite profile, containing the wavelengths measured by both systems, is obtained by superimposing the ASY profile onto that measured by the SYM system.

10. The computer, peripherals and power supply system required to operate the profilometer are housed in the towing vehicle which, in the commercial version of the machine, is a van (see Fig. 3). The road surface displacements measured by the sensors, together with horizontal distance data provided by a device fitted to the gearbox of the towing vehicle, are fed into a mini-computer which is programmed to carry out the calculations necessary to obtain the profile. The calculated profile data are stored on a dual-drive floppy disc system and can be analysed on site using a suite of programmes that have been developed at the Laboratory. A visual display unit provides a communication link with the computer and results of profile analysis can be presented on the graph plotter, printer or visual display unit as shown in the block diagram of the computer system in Fig. 4. Operation of the machine to measure profiles is through a simple push-button control panel and trials have shown that the profilometer can be operated successfully by people with no previous experience of the system after one day's training in its use.

Fig. 3. High-speed profilometer

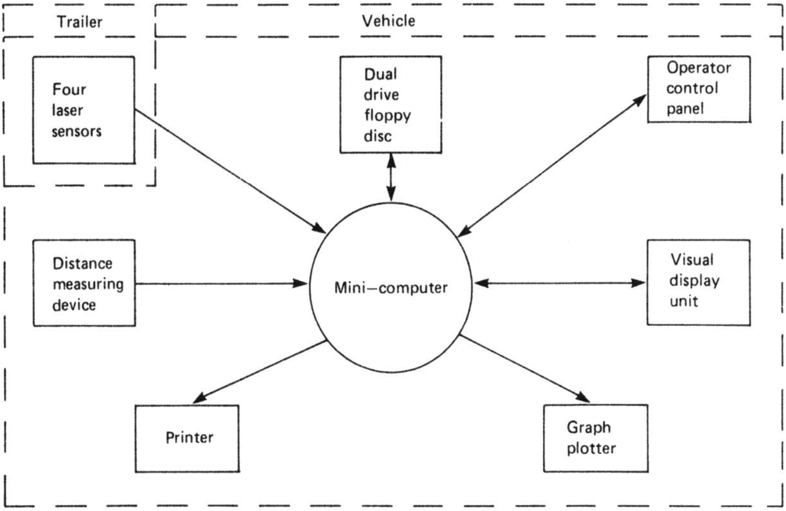

Fig. 4. Computer system of high-speed profilometer

11. Principle of operation. The principle of operation of the profilometer is described in detail in ref. 3 but its main features can be conveniently summarised using the SYM measuring system; the ASY system operates in a similar manner to that of the SYM system. On being moved forward through a distance, ℓ, (the distance between the SYM sensors), the average displacements ($\bar{h}_A$, $\bar{h}_C$ and $\bar{h}_D$) of sensors A, C and D from the road surface are computed. The averaged profile heights, $\bar{Y}$, relative to a datum line defined by joining the centre points of the lengths, ℓ, traversed by sensors A and C, are given by:

$$\bar{Y}_{\frac{\ell}{2}} = 0; \quad \bar{Y}_{\frac{3\ell}{2}} = 0; \quad \bar{Y}_{\frac{5\ell}{2}} = -(\bar{h}_A - 2\bar{h}_C + \bar{h}_D) = \bar{U}_1 \qquad (1)$$

where $\bar{U}_1$ = averaged characteristic measurement of the SYM system.

12. The profilometer is now moved forward a further length, ℓ, such that the new positions, along the surface, of the averaged displacements measured by sensors C and D are coincident with the positions of the previous measurements of sensors A and C. The new averaged profile height $\bar{Y}_{\frac{7\ell}{2}}$ is calculated using the new averaged sensor measurements h'_A, h'_C and h'_D as follows:

$$\bar{Y}_{\frac{7\ell}{2}} = 2\bar{Y}_{\frac{5\ell}{2}} - \bar{Y}_{\frac{3\ell}{2}} - (\bar{h}'_A - 2\bar{h}'_C + \bar{h}'_D)$$

$$= \bar{Y}_{\frac{5\ell}{2}} + \sum_{i=1}^{2} \bar{U}_i \qquad (2)$$

In general, therefore, the averaged profile height $\bar{Y}_x$ at some point $x = (\frac{2n-1}{2})\ell$ metres along the surface is given by:

$$\bar{Y}_x = \bar{Y}_{x-\ell} + \sum_{i=1}^{n-2} \bar{U}_i \qquad (3)$$

13. In the ASY system the spacing of sensors A and B in Fig. 2 is one twentieth of that of the ASY sensors and sensors A and D are common to both the SYM and ASY systems. Using a derivation similar to that given above, it can be shown (ref. 3) that the averaged profile height, $\bar{Y}_x$, of the ASY system at point $x = (\frac{2n-1}{40})\ell$ metres is:

$$\bar{Y}_x = \frac{40}{39}\bar{Y}_{(x-\frac{\ell}{20})} - \frac{1}{39}\bar{Y}_{(x-\ell)} + \bar{W}_x \qquad (4)$$

where $\overline{W}$ is the averaged characteristic measurement of the ASY system. Expressed in terms of the sensor measurements (h_A, h_B and h_D) averaged over lengths $\frac{\ell}{20}$ metres, $\overline{W}$ is given by:

$$\overline{W} = - (\overline{h}_A - \frac{40}{39}\overline{h}_B + \frac{1}{39}\overline{h}_D) \qquad (5)$$

14. Equation 3 defines the longer-wave features of the profile and the short-wave features are defined by Equation 4. A composite profile containing both the long and short-wave features is obtained by superimposing the ASY measurements onto the SYM measurements using a piece-wise linear fitting process (ref. 3).

15. Performance. The performance of the machine has been extensively investigated (ref. 3) and the results are summarised in Fig. 5 in terms of the phase and amplitude responses of the machine, together with the tolerances in measuring profile amplitudes. For wavelengths shorter than 0.3 metres the measured amplitudes are shown to be increasingly attenuated because of the averaging involved in the derivation of the ASY measurements. Phase distortion is shown to be negligible over the range of wavelengths that are of interest for riding quality. Accuracy of amplitude measurement is influenced by both road and machine factors. The maximum tolerances expected for the measurement of wavelengths of 3, 10 and 25 metres are shown in Fig. 5. Test trials (ref. 4) carried out in Germany, in which the profilometer was used over both roads and test tracks, produced results that have confirmed the evaluation results obtained by the Laboratory (ref. 3). Fig. 5 demonstrates the capacity of the new system to measure wavelengths of 100 metres, ie wavelengths well in excess of those that affect riding quality on roads. This is a major improvement on previous systems.

OTHER APPLICATIONS

Measurement of wheel-track rutting

16. Introduction. The development of wheel-track rutting on in-service roads is important because of the implications for road safety and its possible significance as an indicator of the structural state of the pavement. Rut depth is measured in the UK using a straight-edge and wedge; the procedure is slow and hazardous for operators to use on busy roads. To provide a safer and faster method of measurement, the Laboratory has applied the new high-speed profilometer to the measurement of rut depth.

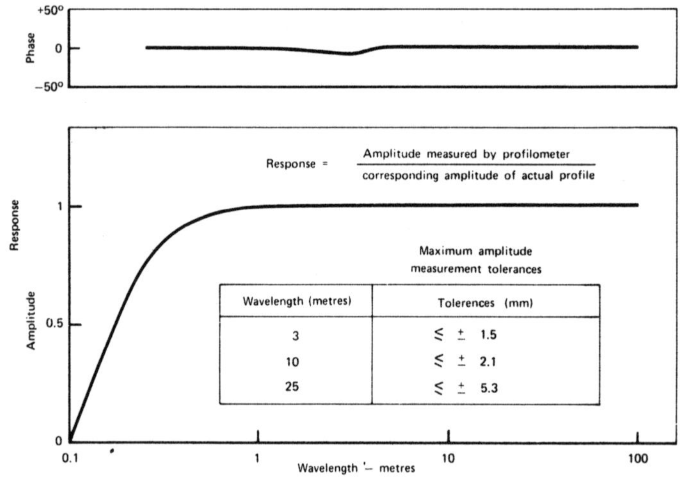

Wavelength (metres)	Tolerences (mm)
3	≤ ± 1.5
10	≤ ± 2.1
25	≤ ± 5.3

Fig. 5. Amplitude and phase response and measurement tolerances of amplitude

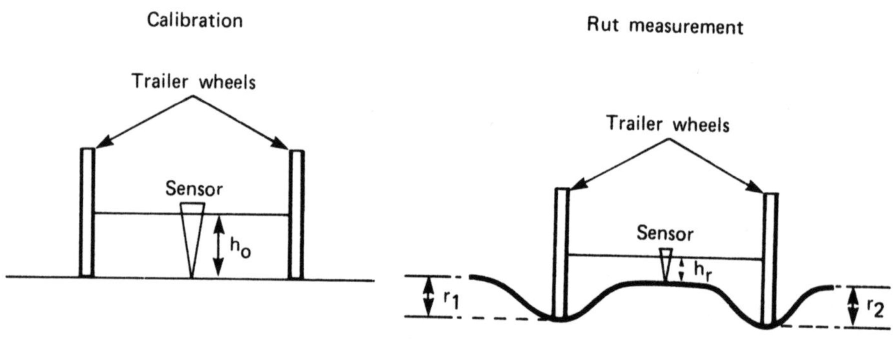

$$h_o - h_r = \frac{r_1 + r_2}{2} = \text{Averaged rut depth}$$

Fig. 6. Measuring wheel track rutting using the high-speed profilometer

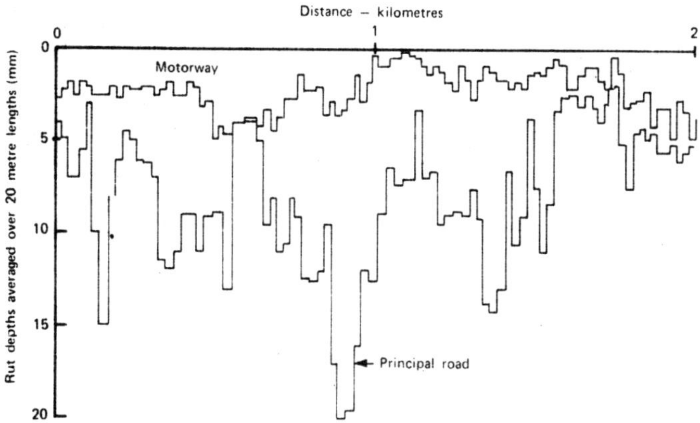

Fig. 7. Variation of wheel track rutting along motorway and principal road

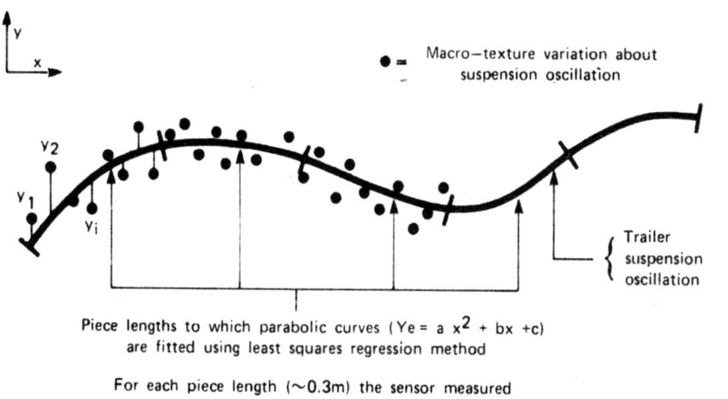

$$SMTD = \sqrt{\sum_{i=1}^{N} \frac{(Yi - Ye)^2}{N}}$$

Fig. 8. Measurement of macro-texture using laser sensor system

17. Principle of measurement. The method used for the measurement of wheel-track rutting is shown in Fig. 6. The wheels of the profilometer trailer ride in the ruts and the laser sensor measures continuously the axle displacement from the road surface along a line centred between the wheel tracks. The difference between the axle displacement measured on a rutted surface and that obtained on a non-rutted surface gives the rut depth averaged over both wheel tracks as shown in Fig. 6.

18. The motion of the trailer on its suspension introduces an oscillatory trend into the rut-depth measurements. To reduce the effect of this trend the measurements are averaged over a length that is related to the speed of the measuring system. For an operating speed of 45 km/h the length chosen is 20 metres; this is increased pro rata with speed. As the measurement is continuous, the averaged value of rut depth obtained is free from sampling error.

19. Performance. Comparison of measurements obtained using the laser system with averaged rut depths derived from straight-edge and wedge measurements have shown agreement to within ±2 mm. The differences between the two methods of measurement arise, in part, from sampling errors associated with the straight-edge system and in part from operator errors in controlling the line of measurement of the laser system. Examples of rut depths obtained operating the laser system, at normal traffic speed, on a motorway and on a principal road are shown in Fig. 7. Facilities available within the system enable lengths of road with critical levels of rutting to be identified. Rut depth data can be stored on floppy disc and analysis results can be presented on the visual display unit, graph plotter and printer peripherals (see Fig. 4).

Measurement of macrotexture

20. Introduction. In the UK, depth of macrotexture, which is of importance for high-speed resistance to skidding, is usually measured by the 'sand patch' method. In this method, texture depth is determined from the area of road surface over which a given volume of sand saturates the texture. The method is labour-intensive, slow and expensive, necessitating the coning off of sections of road to protect the operators. The accuracy and high rate of measurement of the laser has provided a method of making continuous measurement of texture depth at normal traffic speeds. Though only one laser sensor is actually needed

for the measurement of texture, computer programmes have been developed at the Laboratory which permit the use of the high-speed profilometer. More economic single-sensor equipment is also under development at the Laboratory.

21. Principle of measurement. In the displacements measured by the sensor as it moves over the surface, the macrotexture appears in the form of small random variations in the measurements superimposed on the much larger oscillatory trend produced by the motion of the trailer suspension that supports the sensor (Fig. 8). To obtain a measure of the macrotexture depth (ref. 5) the oscillatory trend, generated by the trailer suspension, is first removed using a piece-wise parabolic curve fitting procedure. The texture depth is then calculated in terms of the standard deviation of the residual displacements as shown in Fig. 8. The standard deviation of texture depth, SMTD, is calculated over short lengths of surface, usually of the order of 0.3 metres; it can be presented in the form of a histogram or of a 'profile' of texture depth along the road surface.

22. Performance. Comparisons (ref. 5) of SMTD and 'sand patch' measurements have shown good agreement, with statistically significant coefficients of correlation exceeding 0.9. Though research is continuing into the development of a general relationship between the sand-patch (SP) and SMTD measurements, a relationship based on data from a limited number of test surfaces has been developed (ref. 5) and it is given by:

$$SMTD = 0.12 + 0.59\ SP \qquad (6)$$

Selected lengths of motorway ranging from 6 to 21 km have been measured (ref. 6) using the laser system. The results have shown that the system is capable of continuous measurement of macrotexture at operating speeds of at least 35 km/h (ref. 6). On grooved concrete surfaces the laser system has also been successfully applied (ref. 7) to the measurement of groove dimensions and spacing.

CONCLUSIONS

23. A laser sensor capable of making up to 3300 displacement measurements per second has been developed at the Transport and Road Research Laboratory. The sensor has an accuracy of 0.282 mm over its working range of 72 mm centred at a point some 270 mm from the sensor.

24. The operation of the sensor and its application to the measurement, at normal traffic speeds, of longitudinal road

profiles, wheel-track rutting and surface macrotexture have been described. Each measurement mode is controlled by the mini-computer that is an integral part of the profilometer system. Measurements can be stored on magnetic media for further processing, or analysed on site using the suite of programmes that have been developed for the system.

25. The high-speed profilometer, which also measures wheel-track rutting and macrotexture, provides an efficient, fast and economic tool for both the construction and maintenance engineer.

26. More economic prototype equipment, based on a single sensor, has been developed at the Laboratory for the measurement of macrotexture and rut depth.

ACKNOWLEDGEMENTS

27. The work described in this paper forms part of the research programme of the Transport and Road Research Laboratory and is published by permission of the Director.

REFERENCES

1. STILL P.B. and M.A. WINNETT. Development of a contactless displacement transducer. Department of the Environment, TRRL Report LR 690. Crowthorne, 1975 (Transport and Road Research Laboratory).
2. DICKERSON R.S. and D.G.W. MACE. A high-speed profilometer - preliminary description. Department of the Environment, TRRL Report SR 182 UC. Crowthorne, 1976 (Transport and Road Research Laboratory).
3. STILL P.B. and P.G. JORDAN. Evaluation of the TRRL high-speed profilometer. Department of the Environment, Department of Transport, TRRL Report LR 922. Crowthorne, 1980 (Transport and Road Research Laboratory).
4. BRAUN Dr. Report No. 543. Technische Universitaet Braunschweig 1980 (Braunschweig).
5. COOPER D.R.C. Measurement of road surface texture by a contactless sensor. Department of the Environment, TRRL Laboratory Report LR 639. Crowthorne, 1974 (Transport and Road Research Laboratory).

6. HOSKING J.R. and P.B. STILL. Measurement of the macrotexture of roads PART I. Trials with the contactless sensor on motorway M1. Department of the Environment, Department of Transport, TRRL Report SR 498. Crowthorne, 1979 (Transport and Road Research Laboratory).
7. STILL P.B. Measurement of groove dimensions on concrete pavements by a contactless sensor. Department of the Environment, Department of Transport, TRRL Report SR 443. Crowthorne, 1978 (Transport and Road Research Laboratory).

10

Instrumentation for monitoring the response of pavements to wheel loading

S. F. BROWN, BSc, PhD, MICE, FIHE, Reader in civil engineering, and B. V. BRODRICK, MPhil, Senior Experimental Officer, University of Nottingham

SYNOPSIS. After a brief discussion of the need for in situ measurements of the stresses and strains induced in highway pavements by wheel loads, available instruments are described. The principles to be observed in taking these measurements and the factors involved in the design of pressure cells for stress measurement are outlined. The importance of instrument calibration is emphasized and techniques for satisfactory installation and monitoring of output are reviewed.

INTRODUCTION

1. Improvements to present empirical design methods for highway pavements are being developed with the aid of structural analysis techniques in an attempt to solve the problems posed by heavy traffic loading and to produce more economic solutions by better utilization of materials.

2. An important aspect of this development is the gathering of data on the stresses and strains set up in pavement structures by moving wheel loads. Absolute values of these parameters, measured at critical points in the structure, are needed for checking theoretical analyses, while relative values from different structures provide a basis, at a more practical level, for comparing one design with another. While vertical deflection, both at the surface (ref. 1) and at various depths (ref. 2) can be measured in completed pavements, instruments for stress and strain measurement can only be reliably used in new construction.

3. The data produced by stress and strain transducers

Sensors in highway and civil engineering. ICE, London, 1981, 118-129

needs to be accompanied by details of wheel loading when structural analysis is being verified. The measurement of wheel loading is the subject of another paper to this conference (ref. 3). Useful observations can be obtained by the application of a known wheel load such as happens during conventional surface deflection surveys.

STRESS MEASUREMENT

Requirements

4. The most important stress measurement and probably the easiest to effect, is vertical stress in the subgrade at or just below formation level. This parameter gives an overall idea of the load spreading ability of the overlying structure and is particularly useful in comparing one pavement with another over the same subgrade. If it is accompanied by horizontal stress measurement, then the more important parameter, shear stress, can be deduced.

5. While most of the applications of stress measuring instruments have been in fine grained soils, some work has also been done on their use in bituminous materials and unbound aggregates (refs 4,5).

General Principles

6. The design and performance of earth pressure cells, or stress gauges as they are sometimes called, has received more attention than any other field instrumentation of interest to highway engineers. The determination of in situ stress, however, remains a difficult problem not least because it cannot be measured directly and must rely on a measurement of strain or deformation within the instrument using an appropriate transducer. Because of the many difficulties involved, accuracies better than about 20% cannot be expected.

7. All earth pressure cells incorporate a diaphragm in contact with the soil. The pressure exerted on the diaphragm by the soil is then evaluated in a variety of ways depending principally on the particular practical application. A typical, simple type of cell is shown in Fig. 1. It has a diaphragm which is built-in to an outer ring and is free to deflect under the action of soil pressure. The strain on the inside of the diaphragm, or its deflection,

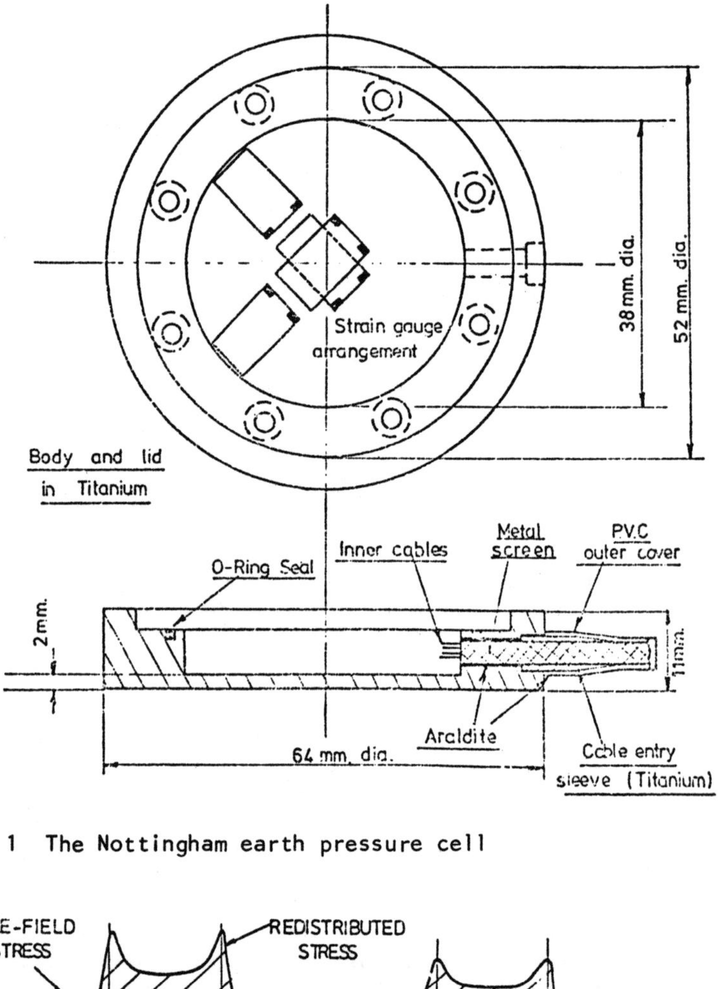

Fig. 1 The Nottingham earth pressure cell

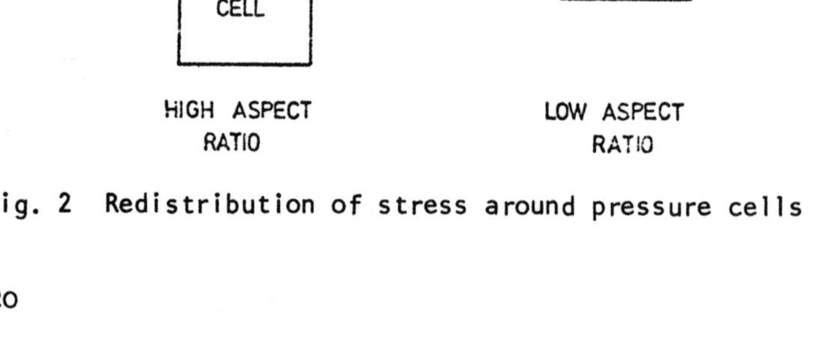

Fig. 2 Redistribution of stress around pressure cells

is then measured using a suitable transducer. Other pressure cells have diaphragms which are restrained by mercury or oil contained within the cell and transmitting the pressure to a second internal diaphragm carrying strain gauges.

8. The introduction of a measuring instrument into a soil mass disturbs the stress distribution as indicated in Fig. 2(ref. 6). The overall aim in designing pressure cells is to obtain a measure of the free-field stress, being the value which would have occurred at the location of the instrument had it not been present. This can best be achieved by designing an instrument which has a minimum disturbing effect though, as in most design problems, this involves certain compromises because of conflicting considerations.

Design Considerations

9. Detailed procedures for the design of pressure cells like that in Fig. 1 to operate in materials of known stiffness are described elsewhere (ref. 7). The parameters of major importance include:

(a) The maximum particle size of the soil which should not exceed 1/50 of the diaphragm diameter.
(b) The ratio of cell thickness to diameter.
(c) The ratio of cell diaphragm thickness to its diameter.
(d) The elastic properties of the soil into which is will be placed.
(e) The material used to construct the cell. Because of the need for preventing corrosion, titanium, stainless steel or an aluminium alloy are usually used.
(f) The type of strain gauge to be used on the diaphragm. Here the basic choice is between a conventional foil gauge with a gauge factor of about 2 and a semiconductor gauge with a factor of about 140.
(g) The maximum stress to which the cell will be subjected during calibration, installation and subsequent operation.

10. Fig. 1 shows that the diaphragm is built into a substantial ring which serves two purposes additional to providing bending restraint to the diaphragm. To avoid the peak stresses shown in Fig. 2 near the edge of the cell, the active diaphragm should not occupy more than 45% of the

total area of the face. In addition, to minimise the sensitivity of the diaphragm to stresses parallel with it, a stiff retaining ring is required.

11. Two other practical design points which require emphasis are the need for good waterproofing and a robust cable entry to avoid damage during installation.

12. The choice of strain gauge type or other transducer element will depend on the monitoring equipment to be used. A compromise has to be reached between diaphragm stiffness, a high value being desirable, and adequate electrical output from the transducer which requires the opposite.

Types of Instrument

13. A wide range of instruments has been produced for in situ stress measurements and details of a range of these, intended for various applications, are given elsewhere (ref. 7).

14. The simplest type of cell is the "free diaphragm" design typified by the instrument shown in Fig. 1. Alternative transducer elements may be foil or wire strain gauges, semi-conductor strain gauges for greater sensitivity, or small linear variable differential transformers (LVDT's) measuring the central diaphragm deflection. Those cells with indirect diaphragms use similar transducers.

15. The piezoelectric cell used by the Transport and Road Research Laboratory contains quartz crystals sandwiched between two diaphragms with copper shims collecting the charge generated by applied dynamic pressure (ref. 8).

Monitoring Equipment

16. A pressure cell of the type shown in Fig. 1, with a foil strain gauge bridge will produce an output of about 0.5 μV/kPa/V. It is desirable that such a small output should be amplified and this may be done with either an AC or a DC system.

17. Most pressure cell installations, both on site and in the laboratory, have used analogue recorders to monitor instrument output when measuring transient stresses. The analogue signal then has to be digitized for subsequent

processing. This procedure can be simplified by use of amplifiers on each cell circuit ensuring a common calibration and easier interpretation of the analogue traces. Amplified signals may also be recorded on magnetic tape, provided reliable software is available to process the digitized signals on a computer. Another alternative is the use of peak hold circuitry which, with either a printer or punched tape output, provides the readings quickly and in a convenient form. If large numbers of readings are required, then microprocessors could probably be used but no experience has yet been gained in this sphere.

Calibration

18. Fig. 2 indicates that a pressure cell will, in general, not exactly register the field stress, though with well designed instruments the error may not be very large.

19. Calibration experiments (ref. 9) with two different cells in two soils under a variety of stress conditions, showed much greater variation in cell registration than was predicted from theoretical considerations. However, when six or more results for a particular soil and cell combination were averaged, the result agreed well with the theoretical prediction. This variability, even under well controlled conditions, was considered to arise from effects associated with placement technique. Hence, under field conditions, even with experienced personnel, these installation effects are likely to be greater and hence as many duplicate instruments as possible should be used if reasonable accuracy is required.

20. It is thus clearly desirable to calibrate pressure cells under controlled laboratory conditions which reproduce, as closely as possible, the field situation. Two types of test have been used, both employing large diameter (225 mm to 950 mm) cylindrical samples with the cell installed centrally. The samples have been either tested in a rigid cylindrical container reproducing zero lateral strain conditions (ref. 10) or as triaxial samples when the vertical and lateral stresses can be varied in different combinations (ref. 11).

Installation Technique

21. A pressure cell orientated to measure vertical stress

is installed "diaphragm-up" in a shallow precut hole in the material surface. For fine grained soils the hole may be cut with a purpose made tool. A trench is provided to accommodate the cable and material is carefully backfilled and compacted over the diaphragm. Care should be taken to remove any large particles (greater than 1/50 of the diaphragm diameter) from the material to be placed in contact with the diaphragm. Techniques have also been developed for installing cells to measure horizontal stress (ref. 12). In coarse granular materials, the instrument may be "prepacked" in polythene, retaining the finer particles in contact with the diaphragm (ref. 5).

STRAIN MEASUREMENT

Requirements

22. There are two components of strain which require measurement in pavements. The transient value, which is recoverable and normally termed "resilient", is important in developing theoretical models for reponse to individual wheel loads while the residual or permanent strain in the vertical direction, when summed with depth gives a measure of the permanent deformation of the surface. Variation in this parmaeter with transverse distance leads to the formation of a surface rut. The depth of the rut beneath a 1.8 m straight edge is the major measure of pavement serviceability in the UK.

23. One of the parameters used in the structural design of flexible pavements using theoretical concepts is the maximum tensile strain in the bituminous layer. This is the criterion for judging the possibility of failure by fatigue cracking. Data is, therefore, needed on the spectra of tensile strains experienced by bituminous layers in service, as well as spot values induced by known wheel loads.

24. The maximum resilient vertical strain in the subgrade is also used in design calculations, but in a less fundamentally based procedure. However, its measurement in situ would provide further useful data for design purposes.

25. Relatively little information is available concerning the contribution of the various pavement layers to the

surface permanent deformation. Measurements of the variation in vertical permanent strain with depth are of use in this context.

General Principles

26. An in situ strain measuring device should be such that it does not reinforce the material in any way. It should be free to deform with its surroundings. Relatively stiff devices which rely on the direct deformation of an electrical resistance strain gauge can only be used for correspondingly stiff materials such as lean concrete or bituminous materials under moderate temperature conditions. For soils and unbound granular material, the strain has to be determined from the relative movement of two points in the material at either end of a known gauge length. The various instruments which are available for both applications are discussed in the following sections.

Strain Coils

27. Two flat circular wire wound induction coils are placed, generally coaxially, but they can be used in a co-planar or orthogonal mode. An alternating voltage set up in one induces a voltage in the other, the magnitude of which depends on the coil spacing, but the relationship is non-linear. The coils and the operating equipment are available commercially from Bison Instruments and field use has been reviewed by Selig (ref. 13). An electronic linearising unit has recently been developed to facilitate the taking of transient readings with this equipment (ref. 5).

28. Coil spacing can be up to four coil diameters but two is favoured and various sizes from 25 to 356 mm diameter have been used in a variety of projects. This instrument is suitable either for dynamic or static work and long term stability is good. They have been used in asphaltic materials as well as soil. The principle of operation is illustrated in Fig. 3. For transient measurements, strains down to 0.005% can be resolved and for long term measurements the figure is about 0.1%.

29. The bench calibration of strain coils is carried out in an apparatus wherein the spacing of coils can be determined by a micrometer and accurately related to the electrical signal. If the medium in which the cells are

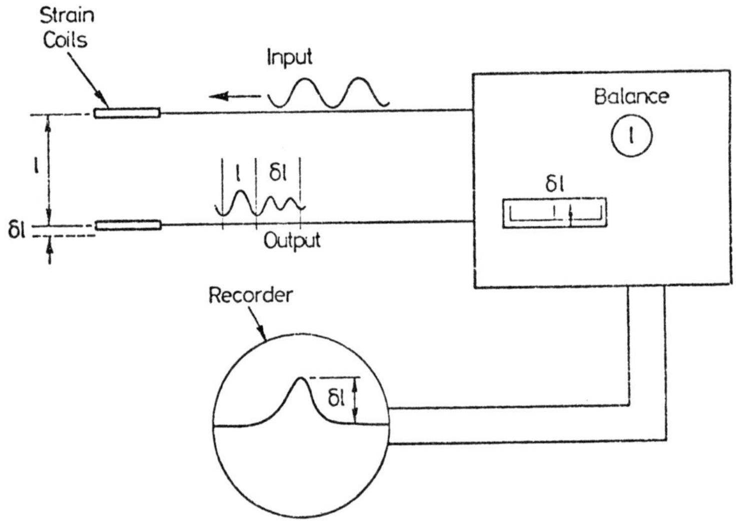

Fig. 3 Principle of strain coil operation

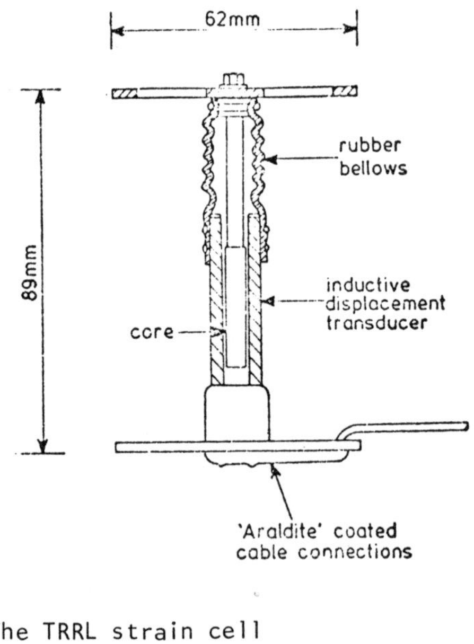

Fig. 4 The TRRL strain cell

to be used is likely to contain ferrous metal ions, special calibration may be required by placing the coils in samples of the material. An indication of the need for this can be obtained by placing a sample of the material between the coils when they are set up in the bench calibration apparatus.

30. The coils are fairly tolerant to errors caused during installation or subsequently which involve movements in directions other than the one in which measurements are being taken (ref. 13). Problems can arise when they are used in the coplanar mode.

31. In order to determine the variation of strains with depth in a pavement, coils can be placed in a stack and their output monitored in pairs.

32. Detailed procedures for installation of strain coils have been described by the Authors elsewhere (ref. 12). In clay, they are placed in a circular hole cut by a special tool. The most convenient procedure is to place the coils at interfaces as construction proceeds, provided the layers are not too thick. If they are, then carrier blocks or mounds of precompacted material may be used to position the instruments. It is important that the stiffness of carrier blocks does not exceed that of the surrounding material, or low readings may be obtained.

33. Similar monitoring equipment to that used for pressure cells is appropriate, but signals do not generally require amplification as a full scale deflection of 5 Volts can be obtained from transient measurements.

Strain Cells

34. Fig. 4 shows an example of a type of instrument which has been successfully used for strain measurement in soils and granular materials, this particular one having been developed by TRRL. A displacement transducer monitors the relative movement of the two end plates. The instrument relies on a satisfactory shear break developing to provide a transition from the uniform deformation in the soil between the end plates and the concentrated movement of the instrument at one end. This is achieved by greasing and particularly by the rubber bellows over the parts where relative movement takes place. The bellows also prevent soil from inhibiting free mechanical movement.

35. The main shortcoming of this instrument is the mechanical linkage between the ends of the gauge length which can cause problems of installation and introduce friction which inhibits free movement, both disadvantages which do not apply to the strain coils. However, the external monitoring equipment is less expensive than for coils and under favourable circumstances smaller strains can be resolved.

CONCLUDING REMARKS

36. The instruments described in this paper have been used in full scale pavement experiments and in laboratory tests. More experience has been gained with the use of pressure cells and these are relatively easy to install in cohesive soils, while techniques have been developed for their use in other paving materials.

37. The use of both stress and strain measuring instruments has been restricted to well controlled research investigations in the past. This is partly because some skill is needed with their installation but also because of the relatively small number of engineers involved in research and development work on pavement design, a situation which is changing only slowly.

38. Field instrumentation is not an exact science so accurate results are not to be expected, even with good practice and ideal site conditions. Replicate measurements should be made whenever possible to produce more reliable means. It is better to have several measurements of one or two parameters from well understood instruments, than a variety of readings. The state of the art in this field is such that equipment cannot readily be bought "off the shelf" and used in a range of circumstances. In this connection, strain coils are more flexible and available than other devices.

39. The importance of adequate calibration should be emphasised. Since any instrument, when installed, will disturb the material in which it is placed, this disturbance needs to be quantified if reliable results are required.

REFERENCES

1. Kennedy, C.K. and Lister, N.W., "Prediction of pavement performance and the design of overlays", TRRL

Report LR 833, 1978.

2. Shackel, B., "The heavy vehicle simulator system in South Africa", Paper presented to the 10th Conf. Aust. Road Research Board, Sydney, 1980.
3. Moore, R., "Measurement of axle loads", Proc. Conf. on Sensors in Highway and Civil Eng., ICE, 1981.
4. Brown, S.F., Bell, C.A. and Brodrick, B.V., "Permanent deformation of flexible pavements", Report to US Army, Univ. of Nottingham, 1977.
5. Brown, S.F., Brodrick, B.V. and Pappin, J.W., "Permanent deformation of flexible pavements", Report to US Army, Univ. of Nottingham, 1980.
6. Selig, E.T., "A review of stress and strain measurement in soils", Symp. on Soil-Structure Ineraction, Univ. of Arizona, 1964, pp 172-186.
7. Brown, S.F., "State-of-the-art report on field instrumentation for pavement experiments", Transp. Res. Record 640, 1978, pp 13-28.
8. Potter, J.F., Mayhew, H.C. and Mayo, A.P., "Instrumentation of the full-scale experiment on A1 trunk road at Connington, Hunts", TRRL Report, LR 296, 1969.
9. Brown, S.F., "The measurement of in situ stress and strain in soils", Field Inst. in Geot. Eng., Butterworths, 1973, pp 38-51.
10. Williams, A.A.B. and Brown, S.F., "The use of earth pressure cells in some road experiments", Proc. 5th Reg. Conf. for Africa on Soil Mech. and Found. Eng., Luanda, Angola, 1971, pp 5.15 - 5.21.
11. Brown, S.F., "The performance of earth pressure cells for use in road research", Civil Eng. and Pub. Wks Rev., Vol. 66, 1971, pp 160-165.
12. Brown, S.F. and Brodrick, B.V., "Stress and strain measurements in flexible pavements", Proc. Conf. on Measurements in Civil Eng., Soc. of Strain Mes. and ICE, Newcastle, 1977.
13. Selig, E.T., "Soil strain measurement using inductance coil method", ASTM Sp. Tech. Pub. 584, 1975, pp 141-158.

11

Continuous measurement and recording of deflections of structures

J. C. LEWIS, BSc(Eng), PhD, Department of Mechanical Engineering, Imperial College of Science and Technology

SYNOPSIS. The application of hydrostatic levels to monitor differential settlement of road bridge supports is described with particular emphasis on the characteristics which influence their long term stability. The main features of a practical installation are outlined and related to the design of a complete system and typical monitoring procedure. Ways are described of up-dating the electronic instrumention and making the system more sophisticated to suit changing requirements.

INTRODUCTION

1. A necessary part of any system of highway maintainance should be the collection of information about the condition of the highway and defects in its associated structures <u>prior</u> to any serious faults arising.

Normally, defects arising from long-term settlement of supports and piers of road bridges are only identified when consequential damage occurs and becomes visible to the inspecting engineer. Hence, it is of value to be able to monitor these structures after the construction phase and to confirm their stability over a given period of changing climatic conditions.

2. In order to detect any likely trends in settling, it is necessary to use more optically-precise levelling techniques than those adopted during the construction stage, which unfortunately is more demanding on skilled manpower resources. Thus, the possibility of attaching sensors to measure changes in level at different parts of a structure could lead to a more cost effective solution to the data monitoring problem.

Sensors incorporated in a hydrostatic level would appear to be the most appropriate way to cope with the base lengths

Sensors in highway and civil engineering. ICE, London, 1981, 130-137

over which displacement measurements are to be recorded and are effective over the range ± 12 mm: (± 0.5 in) and resolution 25 μm: (10^{-3}in) required.

3. In principle, the fixing of liquid reservoirs to the base of the structure at two remote stations, interconnected by a small diameter bore tube, establishes levels which settle to an equi-potential line in the local gravitational field. By arbitrarily defining one station as the datum, differential settlement is monitored by recording the movement of the liquid level relative to the reservoir at the other measuring station.

4. Various applications of hydrostatic levelling, outlined in the literature, emphasise the instantaneous nature of the measurements taken without particular attention being given to the special constraints imposed by monitoring as part of an on- going maintainance programme.

This contribution is intended to focus attention on the parameters which affect the long term stability of such systems and the necessary precautions to be considered at the design stage of a typical installation.

SYSTEM FEATURES

5. The principal features of a complete system are illustrated schematically in Fig. 1. The sensor stage (1) represents a number of hydrostatic levels arranged in the form of a ring linking the stations sequentially at the base of the structure as a fixed installation. This enables the traverse to be closed at the defined datum station providing a check on the differential movements around the base. To determine whole body movement, an additional reservoir would need to be established at a point remote from the structure to provide a datum similar to the monuments used in optical survey techniques.

6. The electronics stage (2), which conditions the signal from the sensors, represents standard equipment available from electronic instrument manufacturers. After selection of equipment to suit the sensors, the choice lies between equipment which may be part of the permanent installation on site or, alternatively, can by transported readily from site-to-site as part of a complete instrumentation package.

7. The data output stage (3) should incorporate.

a) a functional or calibration check on the installed system prior to collecting the data.

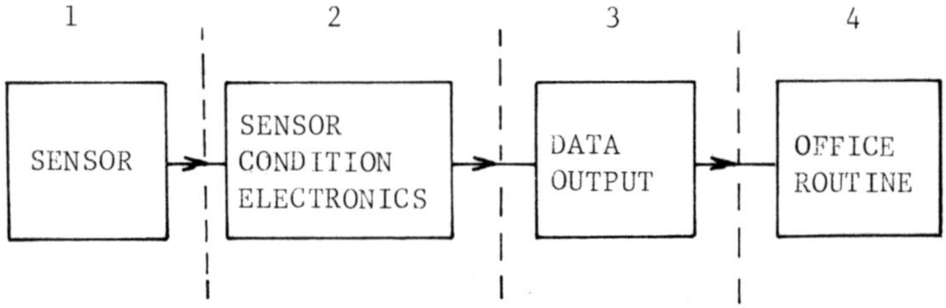

Fig. 1   System schematic

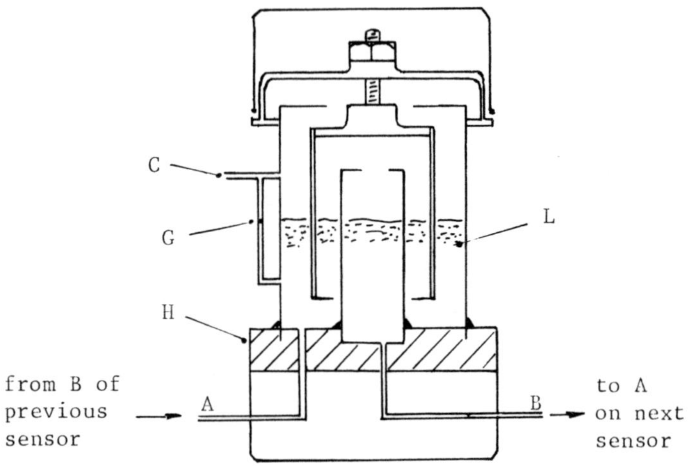

Fig. 2   Hydrostatic level sensor

b) some form of printed output on site as reassurance to the operator that the results being recorded are valid and

c) the actual recording of the data in a form compatible with an office calculator or in house computer.

8. All of this equipment is specified to interface between the technical requirements of the conditioning electronics and to be compatible with planned office procedures for collecting the data, processing in the office and inspection of the trends by the supervising engineer. This is the stage at which the need for action to arrest and prevent further damage is established. It is clear that passing stage 1 to 4 implies not only greater technical complexity of the equipment but also the greater influence that routine office procedures have on the choice of system devised.

SENSOR FOR HYDROSTATIC LEVELS

9. Essentially a hydrostatic level is a simple U-tube manometer where the fluid is perfectly at rest.

When installed with a considerable distance between the arms, settlement of one reservoir relative to the other is monitored by observing the apparent change in level in one arm. Hence, the long term stability depends on two distinct features:

a) the means of measuring change in level of transducers

b) the characteristics of the manometer as affected by the properties of the liquid and the local environmental conditions.

10. Recently, Horsfall[(1)] has described a system incorporating a level transducer in the form of a concentric tubular capacitor, partly immersed in a liquid dielectric. This choice of transducer was mainly justified because of the site problems associated with other types of levels involving distance measurement; servo-driven micrometer, glass float and optical lever, ultrasonic transducer, teflon float with inductive transducer.

Nevertheless, an automatic hydrostatic level of 25 m. span has been constructed at C.E.R.N.[(2)] with a reported precision of ± 0.05 mm over a range of 100 mm with a resolution to 0.01 mm. This incorporates a servo-driven needle which detects surface level by conductivity change.

11. However, capacitive transducers are mechanically rugged, can be constructed with no moving parts and the materials chosen to minimise dimensional changes with temperature variation if need be. In addition, they offer the possiblity of operating in analogue mode as part of an a.c. bridge system or as the sensor in a variable frequency system which is more readily compatible with digital electronics.

12. For the manometer characteristics, pressure difference (AP) between the surface of the two arms is related to the difference in height of the levels (ΔH) by the expression;

$$\Delta P = - \rho \; g \; \Delta H \tag{1}$$

where ρ = density of fluid in level

g = gravitational constant applicable to the installation

Thus the long term stability of change in level depends on:

$$\Delta H = \frac{\Delta P}{\rho . g .} \tag{2}$$

The influence of atmospheric differences between the spaces above the two reservoir levels can be limited by running a second tube parallel to the liquid tube to "equalise" the pressures. From the relation (2) above, variations in fluid density due to compressibility effects and thermal gradients along the interconnecting pipe run, together with variation in the local gravitational constant can have a secondary effect on the long term stability. Sneddon (3,4,5) has thoroughly reviewed these factors but has not referred to changes in volume which can occur in the installation, particularly with loss by evaporation for a single-ended level.

13. These two sets of virtues have been combined in the modular design of reservoir illustrated in Fig. 2 now under development at Imperial College.

Two open tubes are mounted concentrically on a header plate, drilled through its side to allow connection to the inner and outer liquid reservoirs at A and B. The common space above the dielectric liquid (MFL 18) surface is covered by a cap, providing a seal to the outer tube and allowing a pressure tapping to be made to the air space at C. This connection equalises the air pressure between two remote sensors and connects to a sight glass G incorporating a nominal level or filling mark.

A third tube, passing through the cap and electrically isolated from it, forms the second 'plate' of a concentric capacitor and can be adjusted to give a controlled depth of immersion into the liquid. The depth of penetration adjusts the zero-set capacitance and the three nominal diameters are selected to give the appropriate sensitivity for the dielectric liquid chosen. The header can readily be attached to a mounting fixture suited for cemented-in bolt holes or welded stud fixing. Two identical modules filled with different liquids are used side-by-side to form a binary or dual level for temperature compensation as advocated by Sneddon.

An on-off stop cock connected between A and B allows a common level to be established round a ring of such modules. Connection between stations is envisaged using standard 6 mm plastic covered copper tubing inside a 20 mm diameter polythene tube, giving the pressure connection between the liquid surfaces.

14. The first stage of electronics can be incorporated inside a protective cover fitted over the height adjusting screw of the capacitor terminal.

Multi-core cable can also be fed through to give electrical connections to a central control point for ready operator access.

15. Such a module is capable of satisfying essential criteria of a remote installation i.e.

a) each station can be fully protected and made secure against vandals

b) connections between stations can be run in ducts or pipes under the ground or in steel tubes attached to the superstructure

c) no electrical power required in between inspection intervals

d) part of the system can be serviced without losing continuity indata recording if leakage of liquid occurs or in-situ electronics have to be replaced.

SENSOR CONDITIONING ELECTRONICS

16. In analogue systems, capacitance changes are measured in an a.c. bridge configuration, where out-of-balance signals, derived from the local bridge supply oscillator, are demodulated prior to the data output stage. With low cost microelectronics now available, it is feasible to fit each sensor with its own bridge and demodulator electronics and to incorporate a voltage-to-frequency coverter before relaying the signals to a central monitoring point. With this added stage, regulation errors in transmission can be avoided and greater noise immunity achieved over reasonable distances.

17. In practice, an array of sensors can be set and calibrated in the workshop before installation so avoiding the problem of hiring maintainance staff with servicing skills in electronics. Protection against environmental extremes can also be specified for a fixed installation and would not require a local power supply constantly energised. During site vists, portable instrumentation is used to record the state of the structure automatically before moving on to the next installation, and hence a flexible maintainance routine can be easily established. One possible alternative to the analogue system would be to use the sensor as part of the R-C network of the local oscillator and to relay the information directly in the frequency or digital mode but this approach has yet to be explored in detail.

DATA RECORDING STAGE

18. The portable instrument should have the capability of energising the whole installation at each site and to accept standardised input conditions. It would include a multiplexer, sampling in turn each sensor, followed by a frequency - voltage converter before connection to an analogue-digital converter. This provides the output necessary to drive a local digital display giving a functional indication to the operator. Such a system could be organised either on a basis of the operator logging the data manually using a chart recorder or character-punch, or to provide a dedicated microprocessor system that would allow full automatic recording on to a digital magnetic tape cassette recorder. Many such systems with the micro-computer controlling the multiplexer, A-D converter and tape cassette recorder are now under development and in use.

19. However, the main characteristics of such instruments is their ability to accumulate data at extremely high rates. In this application, it is mainly of value in reducing the

time required to monitor a series of structures as long as care is taken to validate the data before recording occurs. The routine processing of the data in the office can then be carried out to provide a format as specified by the engineer.

CONCLUSIONS

20. The application of a modular hydrostatic level to structure settlement has been outlined. Such a system is capable of being added to existing structures as well as new designs with minimum fixed installation costs. Repairs can be effected with minimal interruption to the data records and history of the installation. The portable instrumentation used in gathering data can be used to monitor different sites in the minimum possible time with the aid of automatic recording equipment without causing day-to-day problems in servicing and availability. The system is readily extended to be compatible with existing computer systems.

REFERENCES

1. HORSFALL J.A.C. and KING G.C.P. New geophysical tiltmeter. Nature, 1978, 274, August, 675-676.
2. OLSFORS J. Automatic hydrostatic level, CERN. Technology note, 1979, May.
3. SNEDDON J. Temperature effects in hydrostatic levelling. Survey Review, 1974, XXII, 1974, October, 361-374.
4. SNEDDON J. System temperature of binary liquid hydrostatic levelling. Survey Review, 1975, XXIII, 177, July, 138-143.
5. SNEDDON J. Further generalisation of the hydrostatic models. Survey Review, 1976, XXIII, 182, October, 362-366.

Discussion on Papers 9–11

Mr B.L. PARKER (*Materials Engineer, Surrey County Council*): The Marshall Committee Report (ref. 1) suggested that highway authorities should monitor their highway networks and obtain sophisticated data relating to structural integrity and surface characteristics, all of which should be considered comprehensively before maintenance decisions are made. Unfortunately, it did not say how this was to be achieved. It is essential that all the relevant data can be correlated without a massive manual operation of matching pieces of paper to a map or other visual representation of the highway network. For example, an output from a sensor representing 100 m of highway may be recorded on a piece of paper about 10 cm long: in Surrey to monitor this item - one of the many required by the Marshall Report - would require 30 000 such sheets of paper.

There are about 80 million characters for manual assimilation in the computer output from one season's SCRIM (skidding resistance measurements) - a stack that fills a rack from floor to ceiling. If the engineer is to have any chance of accomplishing that task, the primary need is to bring all data to a common location base (e.g. the highway inventory or network) in a form which can be recognized by the computer. The software can then be arranged so that extraneous detail is suppressed and the essential elements on which decisions are to be taken presented on a VDU.

To achieve this automatic linking to a computerized network, it has long been the practice to insert a numerical key code into the data stream, each key relating to a specific location. Surrey County Council are now making use of a laser mounted on their SCRIM vehicle so that these critical key codes are read and inserted automatically into the other data being collected at high speed. The method is

Fig. 1. Laser and binary coded beacon

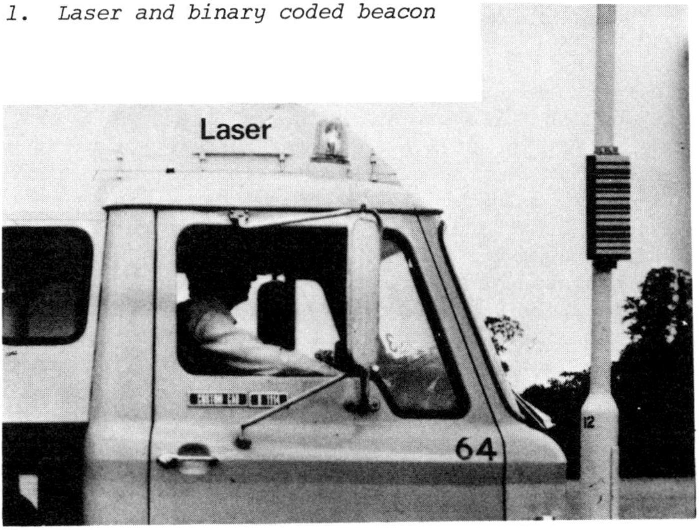

illustrated in Fig. 1 which shows the laser and binary coded beacon. The operator records on a keyboard other essential data relating to the road surface without having to have regard to the exact location of the vehicle.

National guidelines for maintenance rating systems need to be unambiguous; the network should have a common format for main highway features and all raw uncorrected data should be retained in the system so that it can, if necessary, be interrogated. It is essential to be able to link data directly to the highway network; there must be provision for flexible data suppression and computer jargon should be eliminated from the outputs given to the eventual users of the information.

Mr N.W. LISTER (*Head of Highways Department, TRRL*): With regard to the measurement of stress and strains in highway pavements (Paper 10), it is difficult to obtain reliable and accurate results from pavement instrumentation. Aggregates of a maximum size of 40 mm are commonly used and in the upper layers the stress and strain field vary rapidly in relation to distance. The instrument engineer is therefore presented with a very difficult and to an extent almost insoluble problem. If he makes the gauge length of his instrumentation sufficiently long to minimize the distortion of the measured values caused by the presence of large pieces of aggregate, his measurements are likely to be over a length that encompasses an appreciable change in stress or strain. Caution is therefore needed in interpreting the results obtained, particularly when the variability of road materials and the nature of the relations between stress (or strain) parameter and performance are taken into account.

Although it is important to obtain stress and strain information from in-service roads for checking on their performance and therefore on design recommendations, conventional gauges cannot be installed in them. Measurement of deflexion and curvature of the road surface can be made on in-service pavements and has proved valuable, but deductions about associated stress and strain conditions have to be derived indirectly by calculation, and difficulties arise. A more promising and simpler method being developed by the Laboratoire Central des Ponts et Chaussées is to measure the 'ovalisation' of a core-hole in the pavement in response to a passing wheel load.

The in situ strength and stiffness of the sub-grade is more variable than that of any other layer of the road: the top 30 cm of soil is of greatest interest. Has Dr Brown (Paper 10) any information as to the application of the principle of the Maynard pressure-meter (originally developed to study soil strength and stiffness on a larger scale) to the problem of characterizing this important layer of limited depth?

Dr S.F. BROWN (*Paper 10*): Paper 10 discusses the importance of reliable laboratory calibration of instruments in the material to be used on site. Our work with pressure cells in 40 mm crushed limestone wet-mix macadam has involved the successful use of an instrument which is prepacked before installation. Fine aggregate is placed over the sensitive diaphragm and held in place by a thin sheet of plastic (cling film). Details are described in reference 2.

With regard to the French technique involving the distortion of a core-hole in a completed pavement, reference 2 in our Paper describes a similar device developed in South Africa for measuring deflexion at various depths in a pavement.

Some development work has taken place in North America on the use of the pressure-meter principle for measuring sub-grade characteristics. The work is described by Briaud (ref. 3). This procedure still requires further development, but such a device is potentially attractive as part of a pavement evaluation procedure.

Mr P.G. JORDAN (*Paper 9*): With regard to the problem of data collection and use it seems that data are collected either for maintenance assessment or for research purposes. In the case of maintenance assessment only those data required to identify the location and the treatment of defective roads need to be retained; this selective retention of data would reduce the cost of data storage.

For research purposes it is often advantageous to retain all of the data collected over a number of years in order to observe the performance of materials and designs. However, the quantity of data collected for this purpose need not be large as a relatively small and suitably chosen sample of test roads can provide adequate research data.

Mr B.J. THORPE (*Howard Humphreys & Partners*): With regard to the use of the sensor on long lengths of road, using averaging techniques (Paper 9), is there any problem with surface particle orientation under the sensors? Does the uni-directional effect of traffic alter the readings obtained?

Mr JORDAN: TRRL has investigated texture effects theoretically, assuming that the orientation of particles in the road surface is random. We have found that the effect of texture predicted by this theory agrees well with that found in practice in profile measurements.

On sections of motorway near to entry and exit points the profilometer, when used as a rut-meter, does occasionally give unusual measurements; at these points the intersection of the different traffic flows can cause a distortion of the rutting pattern.

Mr G.F. SALT (*Head of Materials Division, TRRL*): TRRL has also been developing the contactless sensor system for measuring texture alone, the aim being to produce a much cheaper instrument than the profilometer. The development has been in two parts: the first has produced a high-speed device using a trailer-mounted sensor towed behind a vehicle carrying a mini-computer, electronic equipment and so on. This system will measure texture for maintenance purposes on highways under traffic. The other part of the development will produce a device that is a miniaturized version of the same thing for measuring texture during construction; this is for use by contractors and local authorities in place of the sand-patch test. Such a machine must be hand-propelled, easily transportable and inexpensive. A prototype is now undergoing trials and it should be available commercially soon.

Mr J. SHREWSBURY (*Transport Planner, Halcrow Fox & Associates*): Would these micro-texture measurements by remote sensors be compatible with, or replace, Scrim measurements, and would they be cheaper and more accurate in doing so?

Mr SALT: They would not replace SCRIM measurements but would be an addition to them, each providing interdependent contributions to an overall measure of skid-resistance.

Dr BROWN: In Fig. 6 of Paper 9 both ruts are equal. When the transverse surface profile of a pavement is measured the near side rut is usually somewhat deeper because vehicles have different axle lengths.

Mr JORDAN: It is generally the case that the near side rut of a lane is deeper than the offside. However, in the development - with time - of rutting there is a high degree of correlation between the near side rut depth and the rut depth averaged over the near side and offside wheel-tracks.

The TRRL rut-meter has been developed to provide a high-speed method of monitoring the development of rutting to detect and locate those sections of the road network where the rutting is critical. The rut depth averaged over both wheel-tracks provides a satisfactory parameter for use with the rut-meter in monitoring the development of rutting.

REFERENCES

1. Ministry of Transport. The Marshall Committee Report on Highway Maintenance, HMSO, London, 1970.

2. Brown, S.F. and Brodrick, B.V. 'Instrumentation for the Nottingham pavement test facility'. Transportation Research Board, Washington, USA. (To be published)

3. Briaud, J.-L. 'Use of a pressuremeter test to predict the modulus and strength of pavement layers.' Transportation Research Board, Washington, USA. (To be published)

12

Inertial platform techniques for rapid surveys of hilliness and bendiness

J. INNES, BSc, MICE, Scottish Development Department

SYNOPSIS. SDD wanted accurate information on Trunk Road alignments and this paper traces the alternative method investigated before settling on Ferranti inertial navigation techniques. The principles involved in inertial navigation and the techniques applied to monitor and compensate for errors are described together with the practical difficulties these imposed on the personnel undertaking the survey at speeds of up to 40 kph. Details are given of the logging and post processing of the data including information on the curve fitting methods adopted. The paper concludes with an indication of the accuracies achieved and the benefits derived from the survey.

INTRODUCTION

1. The Scottish Development Department has been developing over the last few years a computerised highways data base to provide ready access to and manipulation of highway data of various types. This data base consists of 6 definable sub-systems covering Traffic, Accidents, Road Inventory, Road Alignment, Road Conditions and Road Treatments. The sub-systems are related to each other by a common link numbering system which defines uniquely the 600 road lengths making up the 3500km network of Trunk Roads. The Data Base Management System has been developed along with SIA Ltd using SYSTEM 2000 - a soft-ware package marketed by the INTEL Corporation.

Sensors in highway and civil engineering. ICE, London, 1981, 143-154

2. The system continues to develop and now SDD have the following abilities.

a. Allocate the Trunk Road Maintenance monies on an objective basis over the complete network

b. Locate lengths of road with statistically high accident records

c. Automatically produce graphical accident analysis plots which provide the basis for indepth studies of accident sites.

d. Correlate traffic, accident, road conditions and alignments.

One prime objective of the system is to categorise the network on the basis of alignment, traffic flow and road width and ultimately derive a 'capacity index' for each link. Sections carrying flows above an acceptable capacity can be identified and various options simulated to test what improvements are required to extend the road life x years. The identified lengths of road can then be associated with lengths in need of attention from (a) an accident and (b) a maintenance approach to produce a comprehensive, economic and planned programme of works.

3. Therefore SDD arranged to have the data on bends and grades collected by the CHART Survey parties during the maintenance assessment surveys. CHART permits each grade and radius to be assessed in 3 degrees of severity and the Department defined the limits of these categories for use in Scotland. Regrettably the information so collected proved to be unreliable and Ordnance Survey data was insufficient to provide a check. It was considered essential to obtain more accurate data on radius and gradient.

4. Early in 1979 the possibility of using Inertial Navigation techniques become known. Inertial navigation systems had been used in military aircraft, space vehicles and more recently in the North Sea in pipeline surveys in mini-submarines, measuring the movement of oil production platforms under storm conditions and for surveying down oil

wells. There seemed every hope of getting an accurate survey of curvature of gradient at speeds of 40 kph.

DESCRIPTION OF THE INERTIAL NAVIGATION SYSTEM

5. Essentially the Inertial Navigation System (INS) comprises three gyroscopes and three accelerometers. These instruments are held on an orthogonal triad, with one accelerometer and one gyroscope pointing north-south (N/S), another set (E/W) and the third set vertical. This triad, called a cluster is mounted on gimbals. With a motor and synchro on each gimbal, the total arrangement has, with an appropriate servo system, the ability to be gyro-stabilised. This implies that the cluster will remain pointing in one direction irrespective of the motion of the gimbals.

6. At switch-on, an INS must perform two processes. The first is levelling : here, the cluster is aligned to the local vertical. Output signals from the accelerometers become error signals in servos which drive the gimbal motors. Only when N/S and E/W accelerometers detect zero gravitational acceleration is the cluster level. The cluster is levelled to within a few seconds of arc from the horizontal plane. The second process is gyrocompassing alignment - the object is to point the N/S accelerometer to True North. This is effected by measuring a component of the earth's rotation detected by the E/W gyroscope. With a knowledge of the latitude at the point of alignment, this rate is a measure of the misalignment of the cluster from True North. The cluster is then processed until this error is zero. Alignment to North is accurate to within a few minutes of arc. Both these processes are automatic and require about 10 minutes to complete.

7. Having the cluster levelled and aligned to North, the N/S accelerometer detects only accelerations in a N/S direction. Likewise the E/W and vertical accelerometers are pointing E/W and vertical respectively. An acceleration in the N/S direction would be detected by the N/S accelerometer and if this acceleration was integrated, it would produce a velocity in the N/S direction. An integration of this velocity would result in a displacement N/S. Similarly, if accelerations occured in the E/W and

vertical directions, double integration in both channels would produce displacements E/W and vertical.

8. Since it is necessary that the cluster always points North and is aligned to the local vertical, there is a coupling between the N/S accelerometer and the E/W gyroscope. This coupling between the instruments becomes the basis of a feed-back loop and is called a Schuler loop. (See fig I) One property of this loop is that it has a natural period of 84 minutes. (The period of a pendulum having a length equal to the radius of the earth). Thus any instrument errors, scaling errors, computing errors, misalignment errors, etc .... manifest themselves usually as velocity errors whose period is 84 minutes.

9. A typical inertial system used on military aircraft has errors in position of 1 Nautical Mile in one hour. The velocity error associated with the position error has a long period. It is this fact which offers a solution to the problem of attaining an accuracy of centimetres. The solution takes the form of frequently sampling the Schuler loop velocity error. The oscillation can, in short time intervals, be adequately represented by a low order function:

$$a + bt + ct^2 + dt^3$$

where a, b, c and d are constants and t is time. This simple model can then be used to represent the Schuler velocity errors. The accuracy of the model depends on two factors: (a) the frequency of sampling and (b) the accuracy and resolution of velocity error measurement. The first factor is influenced by the mode of operation, and the second factor depends critically on some standard of velocity with which to compare the error velocity. The standard velocity chosen here is zero velocity - stopped. This has a unique quality of being absolute and capable of infinite resolution. In practise, therefore the inertial system has to be stopped at given intervals and the error measured. A model is constructed to represent the error velocity and the net system velocity is inferred from the model velocity and measured velocity.

10. The two variables under the control of the surveyor are, therefore, the length of 'Dwell' period over which the

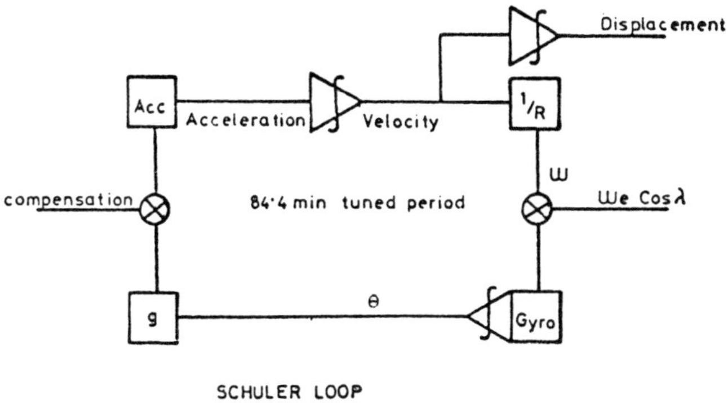

Figure I

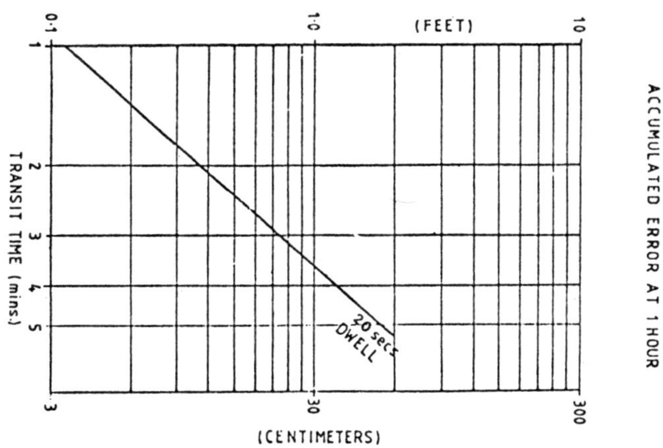

Figure II

velocity update is made and the time in 'Transit' between updates. The minimum 'Dwell' time is that required to filter out the velocity arising from vehicle motion when stopped. A'Dwell' of 5 seconds is sufficient to remove the motion caused by an engine running in the survey vehicle but tests have shown that a 'Dwell' of 20 seconds allows the system to overcome the randomness arising from the inherent noise in the measurement of velocity. No increase in accuracy is achieved by increasing the 'Dwell' beyond 20 seconds unless the vehicle motion 'at rest' is such as to require a longer correlation time. A 20 second 'Dwell' enables the system to accurately identify the velocity error at time 't' from the start of the survey and based on this fig II illustrates the accumulated error after 1 hours' survey for different 'Transit' times between updates.

11. A photograph of the equipment employed to obtain the results is shown in Fig III. The INS is fully contained in one unit measuring 43 x 33 x 25 (cms) and weighing 25kgm. Also shown is the Central and Display Unit (CDU) measuring 14 x 15 x 15 (cms) and weighing 2kgm which as its name suggests is the device for communicating with the IN. The CDU is normally remote from the actual IN box.

PRACTICAL APPLICATION

12. Before commissioning the survey Ferranti undertook road trials to demonstrate the equipment capability and to determine the optimum operating techniques. The INS was mounted in a landrover and driven over a 10 km length of road at different speeds using different lengths of Transit times but maintaining the 20/30 sec dwell period. Unlike space and aircraft applications where the equipment samples at a rate of 2000 per second for this application the sample rate was reduced to one per second which at 40 kph gives approximately 10-11 m between readings. Because the readings are time not distance dependant the equipment automatically decreased the spacing between readings on significant bends and grades as speeds dropped due to the restraints the alignment placed on the driver. Conversely on motorway type alignments the speeds and therefore the sample spacings increased. The equipment surveys and records, relative to an arbitary 0,0,0 starting point, the

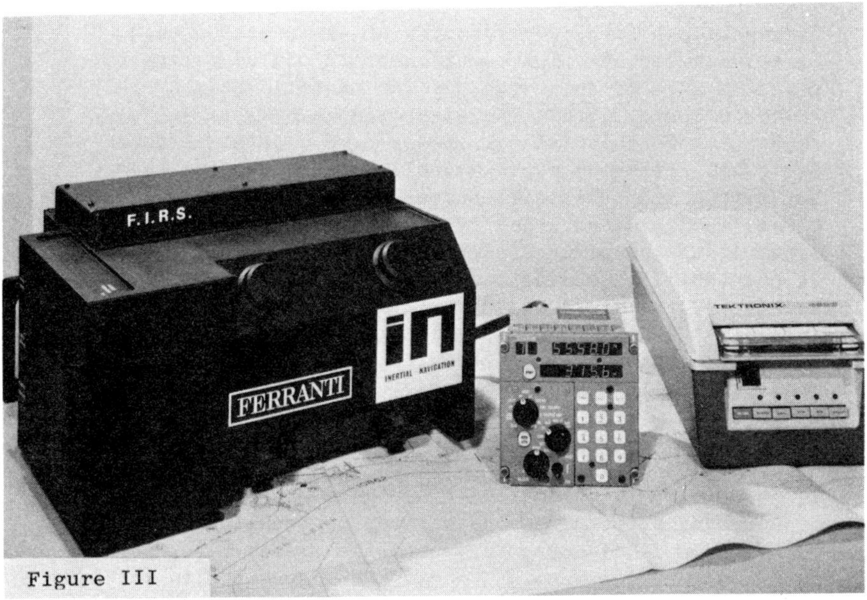

Figure III

Figure IV

3 dimensional polar co-ordinates of its position each subsequent second. The co-ordinates produced relate to the rotating axis of the earth not the magnetic axis or the Ordnance Survey grid. The significance of this point cannot be over-emphasised. Polar co-ordinates at these latitudes differ by approximately 1½ minutes per degree. The output from the trial survey using 3-4 minute Transit times, when plotted, clearly distinguished between the outward and inward legs of the survey. The separate paths of the vehicle could be measured to be approximately 3m apart (1/2 the width of the carriageway) which for this survey was a very acceptable accuracy. Based on this evidence it was accepted that the equipment had the capability to produce the survey.

13. An SDD requirement was that all node points on the network should, if at all possible, be used for Dwell stops in order to relate bend and grade data as a length measurement from these node points in a similar way to the CHART surveys. In many instances the survey party found they could travel from one node to the next within the 3/4 minute transit time without intermediate stops. Stopping at these nodes caused some practical difficulties for, unlike the other Dwell stops required which could be chosen at the drivers discretion, these node points occurred at major junctions. Although the vehicle was equipped with flashing lights and a large surveying sign it still proved hazardous to stop at or even adjacent to those junction nodes due to the volume of traffic. It was decided therefore for operator safety and the convenience of other road users to switch to night-time surveys.

14. Night-time surveys simplified the post processing by reducing the anomalies created by overtaking parked cars and slower moving vehicles or temporary roadworks. Where these anomalies were encountered the operator activated an event key thus raising a flag on the data file and simultaneously logged the nature of the event. A written log was kept for each survey. Navigation at night on unfamiliar roads to locate unmarked nodes at 40 kph from 1 in 10560 scale plans brought its own problems but these proved of little consequence to the results. The vehicle used in the survey is shown on the left of the photograph figure IV. The Bedford van shown on the right has been specially modified for undertaking future survey work of this type. This vehicle will also be used for future

development work which is currently under evaluation.

DATA HANDLING AND PROCESSING

15. The device used for logging the survey data was of necessity rather than choice a papertape punch. The possibility of using RAM store was investigated but rejected because of the constraints the method of dumping the data placed on the operators when working in remote areas. Cassette disc was considered but the equipment available was unsuitable for use in a landrover. The desired rugged magnetic tape unit could not be obtained in time to undertake the survey during the summer months therefore the available and easily maintainable papertape punch was employed. Taking readings at one per second the punch was capable of operating in real time. The transfer of data from papertape to floppy disc and the subsequent processing has been undertaken on a Ferranti mini-computer.

16. Given that a bend was defined as having a radius of less than 610 m and a grade as having a gradient in excess of 4% the following methods of isolating bends and grades and calculating their value were adopted.

Bends. The existence of a bend was established by calculating the spatial rate of change of azimuth with respect to the distance travelled along the road. If the absolute value of this rate of change is greater than $0.23^{o}/m$ then the two points concerned are considered to be on a bend with the sign determining whether it is left or right handed. All adjacent, like signed, points are considered to constitute a bend of radius X where X is calculated by fitting the best least squares quadratic through all these points. The start, end and mid points of the quadratic define a circle whose radius is accepted as the radius of the bend.

Grades. Similarily whether or not two points lie on a grade is decided by checking if the gradient between adjacent readings exceeds 0.038. A set of adjacent points of similar sign are considered to constitute a gradient and the best straight line correlating height to distance is derived.

FIX NUMBER = 10485/75

49.20	0	13382.49	16210.50	24.08	73.47	-.047	.02	2057		
49.27	0	13382.49	16210.50	24.08	73.58	-.047	.02	2057		
49.82	0	13382.67	16211.17	24.07	73.51	-.047	.02	2057		
49.84	0	13383.10	16212.83	24.00	73.38	-.047	.02	2057		DOWN
49.86	0	13383.84	16215.64	23.89	73.36	-.039	-.01	2058		DOWN
49.87	0	13384.83	16219.33	23.72	73.25	-.044	-.03	2059	LEFT	DOWN
49.89	0	13386.03	16223.79	23.50	72.29	-.048	-.07	2060	LEFT	DOWN
49.91	0	13387.37	16223.74	23.26	72.74	-.048	-.03	2061		DOWN
49.92	0	13388.82	16234.10	23.00	72.59	-.046	-.03	2062		DOWN
49.94	0	13390.40	16240.03	22.72	72.94	-.046	.06	2063	RIGHT	DOWN
49.96	0	13392.12	16246.63	22.39	73.51	-.047	.08	2064	RIGHT	DOWN
49.97	0	13394.01	16253.94	22.02	73.36	-.049	-.02	2065		DOWN
49.99	0	13395.99	16261.55	21.65	73.29	-.048	-.01	2066		DOWN
50.01	0	13398.00	16269.34	21.26	73.49	-.048	.02	2067		DOWN
50.02	0	13400.07	16277.34	20.79	73.69	-.056	.02	2068	LEFT	DOWN
50.04	0	13402.31	16295.73	20.26	72.81	-.062	-.01	2069	LEFT	DOWN
50.06	0	13404.73	16294.75	19.70	72.85	-.060	.00	2070		DOWN
50.07	0	13407.30	16304.51	19.14	73.03	-.055	.02	2071		DOWN
50.09	0	13410.03	16314.96	18.61	73.29	-.049	.02	2072		DOWN
50.11	0	13412.85	16325.83	19.15	73.51	-.041	.02	2073		DOWNEND

a. **Basic IN Output**

b. **Edited Polar Co-ordinates**

11526/05

00001			
00002	.31	-1.91	-.03
00003	.75	-4.61	-.05
00004	1.39	-8.38	-.05
00005	2.22	-13.20	-.04
00006	3.15	-18.83	-.05
00007	4.05	-24.68	-.08
00008	5.00	-30.69	-.12
00009	6.14	-37.31	-.18
00010	7.48	-44.76	-.29
00011	8.96	-53.04	-.39
00012	10.54	-62.09	-.48
00013	12.17	-71.82	-.54
00014	13.86	-82.12	-.57
00015	15.52	-92.53	-.56
00016	17.11	-102.68	-.53
00017	18.73	-112.82	-.51
00018	20.49	-123.29	-.47
00019	22.39	-134.12	-.43
00020	24.33	-145.32	-.36
00021	26.29	-156.85	-.27
00022	28.28	-168.69	-.15
00023	30.28	-180.75	-.00
00024	32.30	-192.71	.15
00025	34.38	-204.47	.18
00026	36.58	-216.31	.04
00027	38.84	-228.46	-.17
00028	40.98	-240.99	-.40
00029	42.94	-253.78	-.64
00030	44.71	-266.71	-.88
00031	46.27	-279.68	-1.03
00032	47.64	-292.62	-1.01
00033	48.69	-305.53	-.82
00034	49.36	-318.40	-.44
00035	49.80	-331.21	.00
00036	50.08	-343.95	.46
00037	50.21	-356.64	.92
00038	50.22	-369.32	1.29
00039	50.17	-382.06	1.59
00040	50.09	-394.91	1.85
00041	49.95	-407.90	2.05
00042	49.74	-421.06	2.18
00043	49.50	-434.42	2.25
00044	49.32	-447.98	2.25
00045	49.32	-461.65	2.19
00046	49.72	-475.30	2.11
00047	50.60	-488.88	1.94
00048	52.00	-502.48	1.71
00049	54.09	-516.13	1.48
00050	56.94	-529.79	1.26
00051	60.16	-543.48	1.05
00052	63.31	-557.17	86

c. **Operator Input**

00001	1
00002	2683
00003	11526 5 1
00004	L 26 34
00005	R 44 50
00006	L 54 67
00007	R 71 76
00008	L 77 84
00009	R 85 93
00010	R 121 127
00011	R 185 191
00012	U 203 206
00013	R 207 215
00014	L 219 239
00015	S 45 250
00016	L 262 277
00017	U 268 280
00018	R 278 313
00019	L 314 322
00020	R 338 343
00021	L 345 355
00022	L 368 379
00023	R 382 400
00024	R 422 426
00025	R 439 445
00026	L 450 480
00027	R 481 505
00028	S -1 11528 5 516
00029	R 516 536
00030	S 10 548
00031	L 566 578
00032	R 592 606
00033	L 643 650
00034	R 654 661
00035	R 707 713
00036	R 715 720
00037	L 728 743
00038	L 755 765
00039	R 766 779
00040	S 43 804
00041	L 856 881
00042	R 943 960
00043	L 1087 1099
00044	R 1116 1143
00045	L 1153 1170
00046	S 92 1200
00047	L 1224 1228
00048	S -1 11530 5 1231
00049	R 1245 1252
00050	S 6 1263
00051	L 1285 1300
00052	R 1306 1330
00053	R 1407 1437

Figure V

17. Figure V shows 3 of the outputs produced during the processing stage.

> Printout (a) illustrates the basic IN output with the addition of the initial assessment from point to point of bends and grades. The first 3 lines, which are a sample from some 20, show basically identical readings as the vehicle is performing a Dwell stop.
>
> Printout (b) shows edited polar co-ordinates the Dwell repeats having been extracted together with the readings for tilt and heading.
>
> Printout (c) shows the entries the operation mode manually to the processing system from printout (a) to define the starts and stops of bends and grades. This manual operation has since been eliminated.

Another printout produced but not displayed is the final format of the data for transferring to the SYSTEM 2000 Data Base which took the form of

LINK NO SECTION NO CODE ST CHAINAGE VALUE LENGTH

18. At this stage the initial objective of the survey was complete and once again some 2500 scale overlays were drawn to confirm the accuracy of the surveys. SDD now possessed an accurately digitised, 3 dimensional psuedo-centreline of the network but as yet unrelated to OS. There were two possible methods of skewing from True North to OS co-ordinates, (a) calculating the theoretical grid convergence angle (angle between true and grid north) at the latitude of the starting node, rotate the true results through this angle and then scale the results such that the start point has the required OS values (b) calculating the convergence angle by direct comparison between the true and the OS values of points at the start and end of the survey. Method (b) was adopted. Mathematically the scaling and rotation can be considered as follows:

> Let the start and end of the survey, have OS co-ordinates (X1, Y1), (X2, Y2) and let the true north co-ordinates be (E1, N1), (E2, N2). Then the general scaling and rotation is defined as:

$$X_i = X1 + P\ (Ei - E1) - Q\ (Ni - N1)$$
$$Y_i = Y1 + P\ (Ni - N1) + Q\ (Ei - E1)$$

X_i = X2, Y_i = Y2, if N_i = N2 and E_i = N2 then these equations yield solutions for P and Q. Intermediate OS values can then be produced from, the intermediate polar co-ordinates. This method is valid for links up to approximately 15 km in length.

RESULTS AND BENEFITS

19. As results were produced at various stages they were checked manually by comparison with 2500 scale OS plans. As an example of the accuracies achieved, when the final conversion to OS co-ordinates was produced for a 15km length based on one point at the start and another at the end of the survey length the intermediate check points were within 2m of the values scaled from the 2500 plans. Further the radii produced by the analysis compared very acceptably with known radii on the samples chosen. However as already stated the degree of accuracy is dependent on the Transit time chosen for, as experience has proved, the closing errors on triangulation surveys with this type of equipment can be to within centimetres. The customer must therefore determine the accuracy required for his application.

20. In addition to the prime objective of identification of capacity pressure points, the ability to relate distance as measured on the road surface to OS co-ordinates is an example of the benefits derived from the survey. This ability allows accident locations to be defined as a length along a section, in addition to OS co-ordinates, thus allowing comparison with SCRIM results, CHART road defects data and inventory information as well as statistically to determine the more serious accident locations. The digitised centreline also allows the production of a variety of graphical presentations.

The author would like to acknowledge the assistance provided by Dr R A R Tait, Inertial Applications Group Ferranti (Scotland).

13

Land survey and setting out using EDM equipment and electronic processing

C. P. DEETH, PhD, ASCE, Technical Representative, Aga Geotronics,
M. A. GRANT, BSc, MICE, Director, Eclipse Computer Services, and
T. M. LEON, BSc, ARICS, Managing Director, Glen Surveys Ltd

SYNOPSIS. The historical background to the development of EDM is traced in the context of it's effect on the approach of the surveyor to his task. The role of EDM in increasing the efficiency of engineering site surveys is examined, and it is shown how the field data can now be linked through small desk top computers to the design process and the calculation of earthworks quantities. The same systems can generate data for more flexible setting out during construction. The EDM manufacturers have continually improved and developed their instruments to meet the changing needs of the surveyor.

HISTORICAL BACKGROUND

1. The birth of electromagnetic distance measuring (EDM) instruments arose during the 1950's from experiments carried out by Bergstrand in Sweden to determine the speed of light, leading to the development of the AGA Geodimeter, enabling measurement of distances using light waves. At about the same time research by Wadley for the South African Council for Scientific and Industrial Research produced the Tellurometer, using radio waves; modern versions of both are in production today.

2. Until the advent of EDM determination of distance required the use of physical media of known length. Direct measurements were usually made with tapes of linen, steel or invar, according to the accuracy required. Terrain and vegetation imposed severe limitations. Accurate measurements were made by catenary taping; many corrections were necessary. Really accurate measurement required weeks to measure a single base-line.

Sensors in highway and civil engineering. ICE, London, 1981, 155-163

3. Surveyors therefore resorted to indirect measurement for all but the shortest distances. Theodolites had developed to the stage where angular measurements to one second of arc could be made without difficulty. Triangulation networks enabled many distances to be derived from a single base line. But such a network required more stations and covered a wider area than necessary if direct measurement of traverse lines were possible. The latter could be measured by direct taping or by indirect angular methods such as subtense, using a bar of known length, or by tacheometry using a graduated staff. The range and accuracy of all these methods was limited.

4. The development of EDM enabled accurate direct measurement over long distances, simply by transmitting light or radio waves from one end of a line to the other, irrespective of the intervening terrain or vegetation as long as line of sight prevailed.

5. The earliest EDM instruments were virtually laboratory equipment taken into the field. They were heavy and cumbersome, and required generators or car batteries as their power source. Measurement of a single distance took half an hour or more, which seemed fast compared with days or weeks previously, but which can now be done in less than a minute. (ref.1)

6. Commercial and military demand soon produced more portable and rugged instruments but for some time they remained too large to mount on theodolites and they themselves had no angular facility, so that measurement of angles and distances had to be made as two distinct operations, and their use remained somewhat limited. Within a decade Wild had produced the first theodolite mounted infra-red Distomat, enabling simultaneous measurement of angles and distances.

7. Surveyors themselves were slow to realise the full potential of EDM, only partly due to these physical limitations. They tended at first to use EDM to supplement their traditional techniques, instead of using the different methods which EDM made possible. Even when it was realised that simultaneous radial angle and distance measurements allowed rapid and accurate survey of topographical detail by polar co-ordinates, introduction

of this technique was slow because office plotting techniques had not kept pace with field instrumentation. The introduction of manual polar plotters, and subsequently the use of computer controlled automatic plotters overcame this problem.

8. Surveyors had to reappraise not only their field and plotting techniques, but also the effects of observational errors and the methods of computing and adjusting the field data. They were conditioned into thinking that angles could generally be measured more accurately than distances, and that distance errors were directly proportional to the length measured. It is obvious that if one's tape is too short, the error will increase with distance. It is not entirely true to say that the accuracy of an EDM instrument is the same irrespective of the distance measured, but it is often true for practical purposes, and the proportional accuracy thus improves the longer the distance measured. The effects of angular swing are usually far greater than linear errors when using EDM. The low cost and availability of programmable calculators and desk top computers now enables sophisticated adjustment techniques to be applied to the simplest surveys.

9. EDM has today become widely accepted as part of a surveyor's normal field equipment, as much so as the theodolite or level. In fact many younger surveyors have been weaned on EDM to the extent that they tend to forget that "old fashioned" methods can still be effective in some cases.

10. It has become virtually standard practice to use EDM for engineering surveys: for the control framework; for the detail survey whether by ground or aerial survey and for setting-out. Widespread use has been made for some time of computers and automatic plotters for processing the field data, while more recently the field book is being replaced by electronic data recording systems. The so-called "total station" type of instrument is now available, combining the functions of EDM, theodolite, computer and data recorder in a single field unit.

11. Properly used, these developments enable the surveyor to offer a faster, more accurate survey to the engineer, with greater flexibility in integrating the whole survey,

design, setting out and construction process. Wrongly used, the system can produce field techniques which are a slave to the computer.

THE SITE SURVEY

12. EDM equipment has a role to play throughout a civil engineering project. Taking the design and construction of a highway as a typical example, a site survey is required initially. This survey may be carried out by aerial methods, or solely by ground survey techniques, but in both cases there is a requirement for a control framework. At this stage it is common to find that an EDM instrument is used to observe the distances between the permanent ground markers forming the control traverse along the proposed route. The speed and accuracy with which the distances can be measured is one major advantage of EDM. In addition the topographical detail may also be fixed using the EDM and theodolite, with cost savings compared to more traditional techniques. The measured information may be booked in the traditional manner, or as is becoming more common, recorded in a small portable device. This device may be connected directly to the EDM, enabling some or all of the observations to be recorded automatically, thus reducing the possibility of instrument reading and booking errors.

13. In areas where level information is required for earth moving quantities, grids of levels are unfortunately still often specified. These grids are time consuming to construct, and were only designed to provide the engineer or quantity surveyor with a set of figures which he could easily process by hand. They do not model the ground accurately, as changes of profile between the nodes cannot be represented. They are also wasteful in flat or gently sloping areas where the desntiy of the nodes could be reduced. Many steep sided and irregular ground profiles cannot be satisfactorily modelled in this way.

14. The use of EDM, in one operation allows collection of all the necessary information for both the engineering design and the accurate data base for calculating cut/fill economics and final quantities. As the reflector man can intelligently walk the ground, picking up changes of slope and important features whilst widening his walk in flat areas, this type of survey is more accurate and

comprehensive than the traditional grid of levels.

15. In addition to the control framework, the topographical detail, and the ground levels, the EDM can also be used during the site survey to determine accurate co-ordinates of critical features affecting the engineering design. These may include, for example, the alignments of existing highways to be tied into by the proposed new works or proximity points where clearance is critical. This data can be used during the design process to ensure that the new works will fit the site during construction.

THE COMPUTER LINK

16. Field survey data is recorded in the form of angles, distances and remarks on a field sheet or in an electronic notebook, and is not necessarily in a form for easy co-ordination, plotting or quantity calculation. It is therefore advantageous to feed the survey direct to a computer which will create a suitable data base.

17. The evolution of EDM has been accompanied by a comparable metamorphosis in the computer world. Technical calculations that until recently could only be attempted by large main-frame computers can now be handled by small desk top type machines. Such computers are capable of interactive graphics, manipulating large end complex data bases and creating fine quality plotted output. With these machines becoming an economic reality in small offices, calculations that were too cumbersome to prepare for remote processing at a main-frame can now be processed in a "hands on" environment.

18. The output from EDM can be fed directly into a data base, and plans or sections of the site can be obtained at differing scales or orientations. An accurate method of defining the ground is to divide it into a mesh of triangles where the original surveyed points form the vertices of these triangles. Triangles can be prevented from passing through the air or the ground by ensuring that they do not cross features such as banks, ditches and kerbs. The computer can easily cope with the sizes and shapes of triangles even in cases where the variations in these prevent manual calculation.

19. Roads, drainage and other ground related design data is often computerised. This oiutput is not only printed, but stored on a co-ordinated data base, and is already in the form necessary for creating setting out information with low volume input. For example, assuming that the co-ordinates of control stations have been recorded in the data base and that the geometry of a road has been likewise stored, it is possible to create an entire file of setting-out calculations merely by stating the station, the chainage interval and any off-set requirements.

20. Quantity calculations can be made by comparing any two surveys or models of the same area three dimensionally. The need for creating level grids or sections, even for road quantities, now only exists in order to provide the road engineer with a familiar picture. However, engineers quickly adapt to checking their drawings from contour drawings provided by the computer. Hence the entire project's earth quantities can be calculated in a single complete operation.

SETTING OUT

21. Having completed the site survey and evolved a design based on this data it is necessary to transfer the design onto the ground. At this stage a check survey of the original ground markers may be undertaken to ascertain whether any ground movement has taken place. Furthermore the original control stations were sited for convenience in surveying the detail, their positioning may have been restricted by terrain, vegetation or other obstructions, and as the final alignment was not known, they may therefore be destroyed during construction work. Site clearance before construction enables new stations to be located clear of the proposed works but in positions convenient for setting-out. At the same time the design data can be checked in the field to ensure that it fits the site.

22. The ability to generate setting-out schedules of bearings and distances quickly and economically, and the advent of theodolite mounted EDM, have greatly influenced the surveyors' approach to setting-out. He is no longer required to set out initial points from the control and then progress along the centre-line positioning each peg

in relation to the previous one, and possibly carrying forward an error that cannot be detected until that section is complete. The EDM instrument now allows the surveyor the flexibility of positioning many pegs independently, using radial bearings and distances, from a single station. (ref.2) The position of each peg can then be checked by a rapid measurement to it from an adjacent control station. Even if the surveyor still sets up his theodolite on the centre-line itself, each point can still be positioned and checked independently, reducing the possibility of error. Measurement of the distances by EDM significantly increases the speed and accuracy of the setting-out, particularly over difficult terrain.

THE DESIGN OF EDM INSTRUMENTS

23. Just as surveyors have adjusted their field and office techniques to take advantage of the greater speed, accuracy and flexibility of EDM, so have the manufacturers continually improved their instruments to meet the demands of the surveyor; it has been an interactive process.

24. Most EDM instruments for engineering applications are designed to operate in conjunction with a conventional theodolite. This development has only been made possible by modern electronic techniques leading to a reduction in their bulk and weight, and additionally enabling the use of smaller lightweight batteries. The majority of the instruments use an infra-red light beam which is sent from the transmission optics to a remote reflector station, whence it is returned to the receiver optics (ref. 3). Using different frequencies, producing nominally known wavelengths, superimposed onto the "carrier" beam it is possible to measure distances up to 5 km, the result being displayed digitally within 10 seconds to an accuracy of better than $\pm$ (5mm + 5 parts per million of the distance measured), i.e. $\pm$ 10mm in a kilometre. Further development of the EDM instrument has recently led to the so-called "Total Station", which includes an integral theodolite in which the angle measurements are made electronically.

25. Additional features may also be built into the EDM equipment: often a "tracking" mode may be available, where rapid consecutive distance measurements, of slightly

lower accuracy, are taken to a moving reflector. This mode is obviously extremely useful for setting out, and may also be used to provide a means of acquiring detail point information more rapidly than by individual distance measurements.

26. Many instruments allow for the reduction of measured slope distances to the horizontal, using vertical angles determined either by theodolite and input manually to a calculator, or by an internal electronic vertical angle sensor, in which case the reduction can be performed automatically (ref. 4). It can be appreciated that the computation of height difference may also be made from the same information. As nearly all setting out information is presented as bearings and horizontal distances, the automatic reduction mode can reduce field time considerably.

27. Further features are available, such as an audible signal to monitor the alignment of the EDM instrument with its reflecting prism, and future developments may include the introduction of voice communication from the instrument to the reflector, or possibly two-way transmissions. As mentioned previously the connection of a solid state memory device to either a "Total Station", self-reducing or purely slope distance measuring EDM instrument will enable the recording of some or all of the measured values, which can then be transmitted directly to a computer for automatic processing.

CONCLUSIONS

1. Modern lightweight EDM instruments provide a rapid method of distance measurement for both land survey and setting out tasks.

2. Independent setting out and checking of points is facilitated by the use of EDM instruments, and this procedure greatly reduces the possibility of erros occurring.

3. Measurements made with EDM instruments are significantly more accurate than those obtained by more conventional techniques, and may be made with greater speed and convenience.

4. Automatic reduction of measured slope distances enables the setting out of points on uneven terrain to be accomplished in a rapid and direct manner.

5. The direct transmission of information from EDM instruments to electronic recording devices reduces the possibility for errors in booking and transcription.

6. The state of technology of EDMs and small computers has reached the stage where almost all site development survey and setting out can be carried out more accurately and economically by radials than by other land survey methods.

7. The accuracy of the EDM method reduces errors and significantly improves the accuracy of ground modelling and cut/fill calculations.

REFERENCES

1. SMITH J.R. AGA Geodimeter 1947 - 1978.
AGA Geodimeter Publication 1978

2. BARROW A. EDM IN Setting-out A.S.C.E.
Conference on EDMM & Lasers, Their Use and Misuse, Loughborough, April 1980

3. BURNSIDE C.D. Electromagnetic Distance Measurement
Crosby Lockwood, 1971

4. HODGES D.J. AGA Geodimeter 120, A Field Survey System with Automatic Registration of Field Data.
AGA Geodimeter Publication, 1978

14

Traffic data from aerial photography using semi-automatic methods

Linda J. MOUNTAIN, BSc, PhD, MIHE, Lecturer, University of Liverpool

SYNOPSIS. A photographic technique for collecting traffic flow data at urban intersections was developed. The technique involved the collection of data in digital form from small format photography taken from a hovering helicopter. The data was obtained in this form using a co-ordinate reader and processed by means of a computer which, using a suite of three programmes, output information on a range of traffic flow parameters. It was found that, where comprehensive traffic data was required, the technique could provide a simple, accurate and economical method of traffic data collection and could form a viable alternative to conventional ground survey methods.

INTRODUCTION

1. It has been recognised for many years that photographic methods of traffic data collection can overcome many of the problems associated with more conventional methods, particularly regarding the substantial reduction in manpower requirements achieved, the comprehensive nature of the data which may be obtained, and the permanent record of this data which is provided. Photographic methods of studying traffic behaviour have never been widely used commercially, however, mainly because extraction of the data from the photography has proved to be both time consuming and tedious (ref.1). This problem is particularly acute at intersections due to the complex nature of traffic movements at such locations.

2. It is, indeed, data collection at intersections which poses most problems for conventional and photographic techniques alike. A conventional ground survey designed to provide comprehensive data at a major urban intersection would not only require a large number of observers, but would be expensive and limited in the accuracy with which

Sensors in highway and civil engineering. ICE, London, 1981, 164-174

many of the parameters could be determined.

3. The basic objective of the research was thus to develop a simple and economical method for collecting traffic data at urban intersections. A method utilizing the advantages offered by photography while overcoming the problems traditionally associated with its use.

DATA COLLECTION

Photographic coverage

4. The first problem encountered in any type of traffic survey is that of collecting the required information. In the case of photographic surveys it requires the balancing of such factors as the size of the area which must be studied, the time for which the area must be studied, the frequency with which photographs must be taken, and the quality of the photography in relation to the data to be extracted.

5. Study area. The site selected for study in this particular project was a complex double roundabout system near the centre of Bradford, one with five legs, the other with three, giving a total of 36 possible routes through the study area. In order to include any queues on the approaches it was decided that it would be necessary to include a minimum distance of 50m on each approach to the roundabouts and thus the minimum area of coverage necessary was approximately 0.15 x 0.31 km.

6. Photography used. Consideration of the various photographic techniques available led to the conclusion that small format, time lapse photography, taken from a hovering helicopter could offer the best possibilities for providing the required photographic coverage. The use of small format photography considerably reduces the high film cost frequently associated with photographic techniques, while the use of a hovering helicopter helps to solve the problems of obtaining, simultaneously, a sufficiently large field of view and continuous coverage of the study area (ref.2).

7. Filming was carried out using two cameras: the 16mm Vinten Mark III and the 35mm Robot Motor Recorder 24c. The 16mm format had the advantage that the necessary equipment was readily available and it was found that it could provide coverage of an area of up to approximately 0.64 x 0.46 km. The 35mm photography, on the other hand, had the potential to provide coverage of somewhat larger areas and its higher quality facilitated the identification of different classes

of vehicles.

8. Colour reversal films were chosen for use in both cameras since the aid to vehicle tracking afforded by colour justifies the additional costs over black and white (ref.3). In addition, colour reversal films tend to be sharper and less grainy than colour negative films and are cheaper to process than obtaining prints from negative films.

9. Filming was carried out at a rate of one frame per second. Generally a 1s interval has been found adequate for the extraction of all traffic parameters and longer intervals may be possible - longer intervals offering the advantage of reducing the number of frames to be analysed and thereby analysis time and costs (ref.4). In this study analysis could be carried out both using every frame, and missing frames, so that a variety of time lapse intervals could be simulated and compared.

Problems encountered

10. Vibration and drift. It was found that problems due to helicopter vibration could be reduced to an acceptable level simply by the use of a fast shutter speed. A shutter speed of 1/250s was found quite adequate in the case of a turbine engined helicopter.

11. The problems associated with helicopter drift can be minimised by ensuring that each frame of photography covers a rather larger area than the actual study area. Any drift or rotation of the helicopter then merely results in the study area appearing in a slightly different position on the photograph.

12. Ambient conditions and costs. The only insuperable problems associated with data collection were adverse weather and lighting conditions. In addition, the high cost of hiring a helicopter renders the technique economical only where comprehensive data is required (ref.5).

DATA EXTRACTION

Co-ordinate readers

13. The greatest single problem encountered in photographic traffic surveys is, however, not associated with data collection but with the extraction of the data from the photographs. A variety of methods have been tried, spanning the full range from manual to fully automatic (ref.6). Although automatic methods will probably offer the ultimate solution

to the problems encountered in data extraction, the need for further research and development of both hardware and software, together with a considerable reduction in the costs involved, render semi-automatic methods the most feasible solution for the immediate future.

14. Semi-automatic methods are based on the identification of vehicle positions in terms of photographic co-ordinates, expressed relative to any arbitrary pair of rectangular axes. This is achieved by means of machines known as co-ordinate readers. These machines are semi-automatic in that a human operator is required to identify the points of interest and activate the machines, which then automatically output the rectangular co-ordinates of the points of interest on computer recording media such as paper tape. Semi-automatic techniques are simple to apply, less tedious than manual methods, and the equipment required is readily available and inexpensive in comparison with that required for fully automatic data reduction.

15. In this study a D-Mac co-ordinate reader, capable of reading co-ordinates to ± 0.05mm, and the University of Leeds 1906A computer were used. In order to obtain vehicle photograph co-ordinates, the films were projected onto the surface of the co-ordinate reading table (see Fig.1). The cross-hairs on the projected image viewer were placed coincident with the front nearside corner of each vehicle in turn and a footswitch used to record the co-ordinates on paper tape. When the co-ordinates of all vehicles on a particular frame had been recorded in this way the next frame could be projected onto the co-ordinate reading table and the procedure repeated for each frame in turn.

Co-ordinate transformation

16. Photograph co-ordinates obtained using a co-ordinate reader do not make any allowance for the distortions inherent in single photographs, and the axes and origin will not necessarily be the same from frame to frame. It was thus necessary to convert photographic co-ordinates to a common system of co-ordinates - most conveniently ground co-ordinates related to the National Grid. If the photograph co-ordinates of a number of control points are read on each frame of photography, conversion may be accomplished by the use of mathematical transformation techniques. There are two such techniques available, namely linear transformation (ref.7) and projective transformation (ref.8).

17. Both techniques were thoroughly investigated. It was

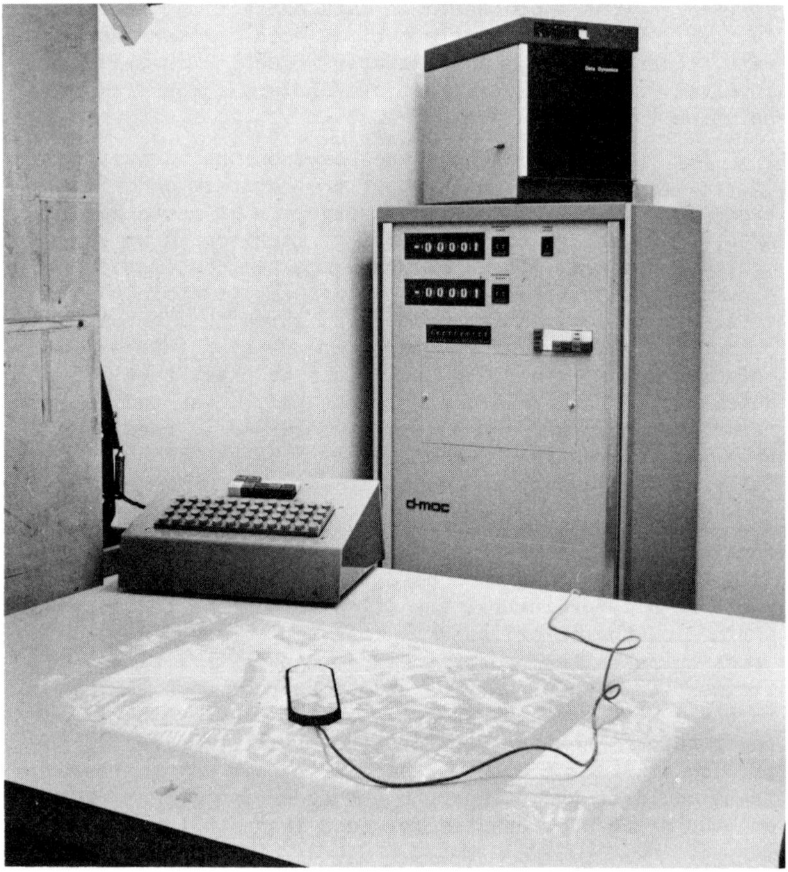
d-moc

found that the conversion of photographic co-ordinates to National Grid co-ordinates could be accomplished quickly and conveniently using either linear or projective transformation techniques, with a level of accuracy adequate for the majority of traffic engineering purposes. While it was found that both methods were capable of similar levels of accuracy in the case of the study area used, projective transformation was preferred for its greater flexibility (ref.6).

Co-ordinate matching

18. In order to obtain traffic data from the photography it was then necessary to devise a technique which would enable the co-ordinates of individual vehicles to be matched from frame to frame and thus allow their paths through the intersection to be traced. Attempts have been made to solve this problem in the past but the techniques devised have, in all cases, been applicable to links where the direction of vehicle travel is constant (ref.7 and 10). A rather different approach is needed for the study of intersections where, not only can a vehicle's direction change, but each vehicle has a choice between a number of alternative directions.

19. Identification codes. The approach adopted was to associate with each set of vehicle co-ordinates, numerical identification codes based on the colour, type and lane of travel of each vehicle. The identification codes were recorded on paper tape, using the supplementary keyboard associated with the co-ordinate reader, prior to recording each pair of co-ordinates (see Fig.1).

20. Zones. Vehicle indentification codes obtained in this way are, of course, by no means unique and it is thus necessary to devise a method by which the mis-matching of the co-ordinates of apparently identical vehicles can be avoided. This was achieved by means of a system of zoning as indicated in Fig.2. The study area was divided into zones such that a vehicle in any particular zone on one frame could only be in one, or perhaps two, other zones on the next frame. In order to follow vehicles, a computer programme was written so that, for a vehicle in a particular zone, a search was made for an identical vehicle on the next frame in the most likely zone first, then in the next most likely zone, and so on until a match was found.

Computer programmes

21. Data obtained using the co-ordinate reader was

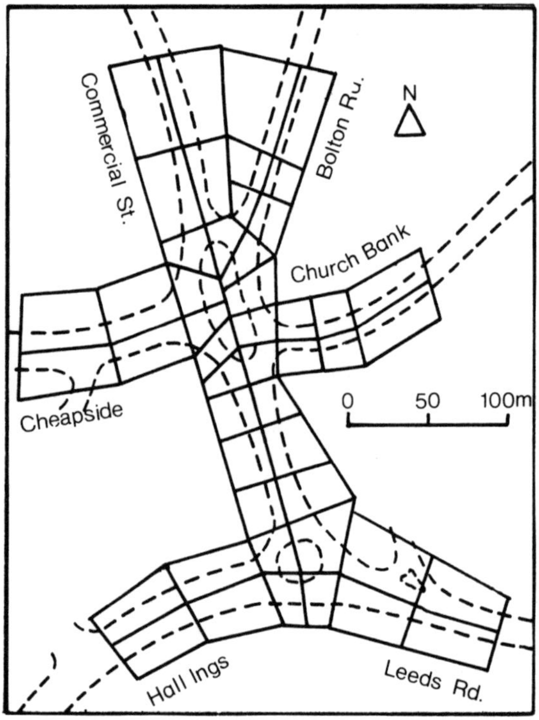

Fig.2. The system of zones used for vehicle tracking.

processed using the ICL 1906A computer and a suite of three programmes. These programmes were designed to transform each set of vehicle photograph co-ordinates to ground co-ordinates and associate with each set of co-ordinates additional identifying codes according to the zone the vehicle was in and the frame it was on. The co-ordinates of each vehicle were then matched so that the position of each vehicle at discrete intervals of time was determined. A number of traffic flow parameters were obtained, including the flow on each leg of the roundabout classified according to direction, the origins and destinations of all vehicles followed through the roundabout system, together with the mean journey time and speed for each route.

Problems encountered in data extraction

22. There were two major problems encountered. Those due to lack of contrast between the road background and certain vehicles, with a consequent possibility of vehicles being missed on some frames, and those associated with vehicles appearing to change colour in areas of varying lighting conditions, frustrating the association of the correct colour code with the correct vehicle. A sequence was incorporated into the co-ordinate matching programmes to take account of these difficulties.

ACCURACY AND COSTS

23. One essential requirement of any useful, new traffic data collection technique is that it should be capable of providing data with levels of accuracy similar to those currently achieved using conventional methods, at similar or reduced costs. In addition to comparing the accuracy of the extracted data, and the costs associated with obtaining this data, using both 16mm and 35mm photography, the efficiency of using 1, 2, 3 and 4 second time lapse intervals was also investigated.

Accuracy

24. In general, it was found that, although the levels of accuracy attainable using the photographic technique tend to be rather lower than those which may be achieved using conventional methods, the accuracy levels were adequate for the majority of traffic engineering purposes (ref.6). For example, it was calculated that, in at least 68.27% of all surveys carried out using the technique, estimated link flows would be within ± 5.8% of actual values, estimated origin and destination flows would be within ± 8.5% of

actual values, and estimated journey times within ± 5.4% of actual values. The level of accuracy was not found to vary significantly with the type of photography used or the time lapse interval.

25. The slightly lower accuracy levels are, in fact, offset to some extent by the additional data which may be readily obtained from the photography. This includes traffic density, vehicle paths and gap acceptance, all of which could not easily be obtained using conventional methods. In addition, ancillary information on external parameters affecting traffic flow may be obtained without recourse to further surveys.

Costs

26. The cost of a traffic survey can be evaluated in terms of two basic units, namely time and money, and it is desirable that the costs in terms of both should be minimised.

27. Time. Investigations showed that an overall reduction in the number of manhours required for data collection and analysis could be achieved using the photographic technique with a 2, 3 or 4 second time lapse interval. The photographic technique, however, requires that the majority of man-hours should be worked consecutively and thus, although the photographic technique requires considerably fewer personnel, the overall time period required to produce traffic data in a useful form tends to be rather longer than that required for conventional surveys. If a 3s or 4s time lapse interval is used, however, an overall time period of under two months should generally be adequate to obtain comprehensive peak hour traffic flow data at an intersection similar to the one studied in this project (assuming that it is not necessary to wait very long for suitable flying conditions). Although the overall time period is about twice as long as the overall time required to obtain similar data from a ground count, the time scale is not excessive and should be acceptable in the majority of cases.

28. Money. Overall survey costs do not vary significantly with either the 16mm or 35mm formats and from this point of view either could be used equally well. The time lapse interval used is, however, one of the major determinants of costs. If it were necessary for the user to pay full agency rates for computer processor time, the photographic technique would be rendered uneconomical irrespective of the time lapse interval used. If it is necessary to pay only overhead costs for computing time, then the use of a 3s or

4s time lapse interval can render the photographic technique an economical method, irrespective of traffic density. The use of a 1s time lapse interval, on the other hand, would make the technique both too lengthy and too expensive.

CONCLUSIONS

29. Small format, time lapse photography taken from a hovering helicopter provides a relatively simple and economical technique for collecting comprehensive traffic data at complex intersections.

30. Extraction of the data from the photography, using a co-ordinate reader to obtain the data in digital form, combined with computer analysis of the data, can form the basis for a simple and relatively rapid data extraction technique. The technique can be used at any location to provide comprehensive traffic data, provided that suitable photographic coverage is available and can provide data with a reasonable level of accuracy at costs similar to or lower than those encountered using conventional ground survey methods.

31. The survey method described can minimize, if not totally eliminate, the data extraction problems commonly associated with photographic traffic surveys. The method is practicable and represents a viable alternative to conventional ground survey methods for collecting comprehensive traffic flow data.

REFERENCES

1. ANON. The practicality of aerial traffic data collection. Institute of Traffic Engineers, Informational Report, 1975, 5-24.
2. GARNER J.B. and L.J. MOUNTAIN. The potential of 16mm and 35mm time lapse photography taken from a helicopter for traffic data collection at complex intersections. Photogrammetric Record, 1978, 9(52), October, 523-535.
3. BRANSTON D.M. A method of collecting data on speeds and headways on a motorway. Traffic Engineering and Control, 1972, 13(10), October, 430-432.
4. DART O.K. Left turn characteristics at signalised intersections on four lane arterial streets. Highway Research Record, 1968, 230, 45-59.
5. MOUNTAIN L.J. Traffic data collection. Traffic Engineering and Control, 1979, 20(3), March, 129-131.
6. MOUNTAIN L.J. The potential of small format photography for traffic data collection, Ph.D. Thesis, University of Leeds, 1979.

7. UREN J. and J.B. GARNER. Integrated system for urban traffic data collection. Transportation Research Record, 1979, 699, 41-49.

8. THOMPSON E.H. Photogrammetry in the restoration of Castle Howard. Photogrammetric Record, 1962, 4(20), 94-119.

9. MOUNTAIN L.J. Urban traffic data collection from small format photography. Proceedings of a Conference on Traffic Data Collection, University of Leeds, 1979, January, 143-166.

10. TAYLOR J.I. Photogrammetric determination of traffic flow parameters. Paper presented at the 11th International Congress for Photogrammetry, Lausanne, Switzerland, July, 1968.

Discussion on Papers 12–14

Mr R.J. POTTER (*Consultant, Wootton, Jeffreys & Partners*): Over the past two years we have used helicopters and aerial photography for collecting traffic data. We overcame the problem of transforming the photographs by employing a photographer to lean out of the helicopter to adjust for the drift and rotation of the helicopter. We found, as did Dr Mountain, that having to pay commercial rates for computer time made automated data-reduction methods uneconomic. The data obtained by this method is very valuable as it provided route information as well as origin-destination information. This technique has only been used for data collection in city centres because of the limited area that can be covered.

Because helicopters are expensive we normally use video recorders on the ground and the analysis is performed on our micro-computer based system called VISTA. Any discrepancies that may appear at the modelling stage can be sorted out by the ability to refer back to traffic conditions at the time of the survey. That has proved a valuable asset.

Mr D.M.A. HOOK (*Partner, G. Maunsell & Partners*): Where do the flying costs appear?

Dr L.J. MOUNTAIN (*Paper 14*): Although, in absolute terms, helicopter hire charges may be considered expensive (the commercial rate being of the order of £200 an hour in 1980), they do, in fact, form a fairly small proportion of the overall costs which would be involved in collecting and analysing traffic data, using either photographic or conventional techniques. In the case of the particular intersection described in Paper 14, for example, it was shown that, provided that it was only necessary to pay overhead costs for computing time, helicopter hire charges represented about 10% of the total costs of the photographic survey (depending on the volume of traffic and time-lapse interval), - compared with the cost of the use of a co-ordinate reader to analyse the data which represents between 45% and 65% of the total. If, on the other hand, it was necessary to pay full agency rates for computing time, then the major portion of the costs would be associated with computer processor time - about 60% of the total cost - compared with 25% for the use of the co-ordinate and 5% for the hire of a helicopter.

It is important to stress that, in comparing the costs of the photographic technique described in Paper 14 and conventional techniques, the photographic technique would prove to be economical only where comprehensive traffic data were required concerning fairly complex traffic movements. If, for example, simple volumetric data alone were required, then terrestrial photography combined with manual analysis, or conventional survey techniques would undoubtedly be cheaper.

Mr J. SHREWSBURY (*Transport Planner, Halcrow Fox & Associates*): To what practical use has Mr Innes put the very detailed data obtained from the inertial navigator? In particular, the SDD planned to work on relating road geometry to accident frequency. Has this been done?

Would Mr Innes comment on the matching of the inertial navigator data which is accurate to about 1 cm, with the accident location data which is good to about 100 m.

Experience shows that visibility is a complex matter, and sight distances cannot always be derived from carriageway geometry (i.e., inertial navigator output) alone. However, the combination of inertial navigation and video recording has great potential in this field. How far has this been researched?

Mr T. BARRETT (*Head of Technical Services Unit, TRRL*): With regard to the photographic technique for collecting traffic flow data (Paper 14), TRRL has considerable experience in this field and has developed a number of techniques. Referring to the problems of vibration and drift in the use of a helicopter platform, we also found that a shutter time of 4 ms is satisfactory for a helicopter with a turbine engine. We have tackled the problems of drift by mounting the camera and a special sight on a gimbal system which enables the operator to keep the camera aligned on the main aspects of the scene. However, the use of a helicopter is expensive at more than £150 an hour and we have examined other techniques.

Fig. 1. A typical record using a fish-eye lens taken from a balloon

Fig. 2. A typical record using a fish-eye lens taken from a mast fitted to a Landrover

Fig. 3. Lamppost camera fitted for time-lapse use

Fig. 4. A record obtained using a lamppost camera

A tethered helium-filled balloon has been used to study traffic flows at a junction. The limitation in the weight of the equipment which can be lifted by a balloon is much more severe than in the case of a helicopter. Like the helicopter, the balloon suffers from restriction by weather conditions: steady winds are acceptable but it is quite unsuited to gusty conditions. A typical record using a fish-eye lens is shown in Fig. 1 - at £150 a day the use of a balloon is well worth considering.

A further cheaper alternative is the use of a road vehicle with an extending mast. A landrover was fitted with a mast which could be extended to 15 m. Using a camera fitted with a fish-eye lens it is possible to cover an area of up to 120 m diameter giving a picture as shown in Fig. 2. The compression of the scale at the edges of the photograph can give difficulties in the assessment of traffic movements but nevertheless the equipment has been successful for many applications and its operating costs are quite low.

Figure 3 shows a lamppost camera fitted for time-lapse use. The oblique view shown in Fig. 4 may not be suitable for many applications where the overall flow round a complex intersection is being examined. The installation is suitable for use in congested streets or at a small congested junction because it can be mounted on a street lighting column and left to operate independently for the duration of its film load. Its use avoids the positioning of a vehicle in a street and the need to gain access to a building giving the view required. Under these circumstances it is cheap to install and operate. In some applications two lamppost camera units have been used to cover a 24 hour period, one filming during the day in colour, the other during the night in monochrome. Switching from the day camera to the night camera and vice versa is effected by an ambient light-sensor device. To avoid frequent loading and reloading of the camera it may be necessary to reduce the picture rate from the normal rate of one every half second to one every 16 seconds.

TRRL generally uses 16 mm film which on an hourly basis is only about a sixth of the cost of 35 mm film but some cameras are dual-gauge and this facility can be useful when greater resolution is required. Colour reversal film is used almost exclusively and for many applications its cost is less than the cost of black and white. In some cases, video equipment may be used.*

DISCUSSION ON PAPERS 12-14

Dr MOUNTAIN: I agree with Mr Barrett's general point that, because of the wide range of possibilities for a camera platform photographic methods of traffic data collection can be readily tailored to the job in hand. However, it is important to note that one particularly important advantage of the use of a helicopter over a tethered balloon is that any local authority, or other potential user, could hire a helicopter fairly readily, whereas balloons are less widely available. Similarly from a practical point of view, although stabilized, gimballed mountings can give excellent results in terms of overcoming the problems of both drift and vibration. Enquiries made of commercial manufacturers of such equipment (Marconi-Elliot Avionics and Wesscam Camera Systems) indicated that the costs of either buying or hiring such equipment would be prohibitively high for traffic surveys.

A fuller review of the types of photography available and the traffic data which may be obtained from it, together with a fairly comprehensive bibliography is contained in a recently published paper 'Applications of photography to traffic surveys' (ref. 1).

Mr GERAGHTY: On video techniques there can be many problems, principally in getting the power to the camera and controlling the power. Has any thought been given to battery-powered cameras?

Mr BARRETT: Thought has been given to this problem but we have not yet done any work with more portable cameras. Good quality, battery-powered lightweight cameras are under development and should be available in the near future.

Mr SHREWSBURY: Does Dr Mountain or Mr Potter know what work is being done on the fully automatic processing of photographs? Current methods need someone to locate, identify and, in some cases, watch vehicles, and this is still a major cost in the data collection.

Dr MOUNTAIN: Attempts have been made to apply automatic image processing to traffic data collection (refs 2 and 3) but the techniques have not as yet been fully investigated. Although they can offer many advantages, and could form a feasible alternative to the other methods, there are a number of drawbacks associated with their use (ref.4). Apart from the cost of the equipment, which is at present prohibitively high, the major problem appears to be the minimization of the computer memory size requirements. A digitized image contains as many as 160 000 numbers per cm^2 and there is thus a need to devise a method which will allow the techniques to be applied with considerably fewer data points.

Mr POTTER: At Wootton Jeffreys & Partners we used the portable version of the VHS format (video recorder). These are all battery-powered and there were no problems.

Fully automatic data analysis appears to be an attractive proposition at first sight but it becomes less attractive when researched. We made a rough-and-ready prototype which worked extremely well with a video recording of the straight part on a motorway. However, scenes at roundabouts and so on tend to merge into different shades of grey.

Mr D.I.M. PUTNAM (*Deputy Regional Controller, South West Region, Department of Transport*): A further problem in the use of helicopters is the difficulty of any pilot keeping over a particular location for any length of time.

Dr MOUNTAIN: Although helicopter drift and rotation mean that successive photographs do not cover exactly the same ground area, this does not in fact produce an insurmountable problem. If the field of view of the camera is sufficiently large, it is possible to ensure that helicopter drift and rotation simply result in the study area appearing in a slightly different position on successive photographs.

Dr S.F. BROWN (*Paper 10*): Have TRRL considered using the laser technique for measuring transient surface deflexion as an alternative to the slow mechanical procedures involving the deflectograph on deflexion beam?

Mr P.G. JORDAN (*Paper 9*): Deflexion is measured in units of hundredths of a millimetre. However, the depth of texture on a road surface is of the order of millimetres. Consequently the TRRL laser sensor with its point measurement and vertical resolution of approximately 0.25 mm is not suitable for measuring deflexion. A more sophisticated sensor is required for this measurement.

Mr M.J.H. CHANDLER (*Head of Urban Traffic Control and Signals, Greater London Council*): The value of automatic data collection is worth reviewing in economic terms for those cities that have centralized control systems for their traffic signals. These cities rely on manually collected data for fixed-time control programmes. The cost of running such schemes is small compared with the value of the traffic benefits obtained.

As the data become out of date, the benefit of the scheme is reduced, and may even become zero, or negative. By this time there may be such a backlog of survey work that it would be difficult to retrieve the situation. It is worth considering whether or not a portable system could be developed which could collect some of the data automatically, perhaps by detectors or a video camera, and suitable software. Clearly there are difficulties, but in relation to the traffic benefits, the cost is not likely to be significant.

Mr GERAGHTY: One of the areas the TCC Division of the Department of Transport have been looking at is the occurrence of accidents and damage on hard shoulders. Records show that 30% of all fatalities on motorways in Northamptonshire in the past three years have been hard shoulder accidents.

REFERENCES

1. Mountain, L.J. and Garner, J.B. Applications of photography to traffic surveys. 'Highw. Engr.', 1980, 27(11), November, 12-19, 27.
2. Baggot, T.A. Traffic data acquisition from aerial photographs by photographic image processing. 'Highway Research Record', 1966, 109, 1-7.
3. Weinburg, H.I. A commentary on traffic surveillance systems and techniques. 'Traffic Engineering', Institute of Traffic Engineers, 1967, 37(b), 24-27.
4. Institute of Traffic Engineers. The practicality of aerial traffic data collection. Institute of Traffic Engineers, Informational Report, 1975, 5-24.

Closing address

R. J. BRIDLE, Controller of Research and Development, Department of Transport, and Director, Transport and Road Research Laboratory

The initiative for this symposium came from the Department of Transport in its desire to generate more data more cheaply, and to stimulate the exploitation of new technology. In all aspects of DoT interests - from ports to highways - engineers face a particular problem: how to sensibly take account of the latest developments that have been triggered off by the advent of micro-electronics. What is going on? What is available? How can it be exploited? How can engineers get optimum use from a system when there are so many overlapping skills and interests in this field? What policy context is there for the development and exploitation of any new proposals?

Micro-electronics can influence policy, design, analysis and management of preparation, manufacture and construction, and control and management once a system is in use. For a relatively low investment abilities are made available which make it possible to get the maximum use out of mechanical, natural and human organizations.

In the 20th century more civil engineering works have been built than ever before in the rest of history. It is evident that infrastructure is being at least imperfectly looked after and it is running down. Often construction projects are over-provided and over-designed, but because too little knowledge and information is available for their design the construction has a built-in obsolescence and reduced utility.

What is needed is knowledge, its dissemination and cheap ways of collecting the necessary information. Information, together with means of its swift organization, manipulation and rapid distillation, is essential to determine optimum investment in a whole life cycle.

We are now passing the stage of the big computer which could not cope with these demands. As a computer becomes larger, proportionately more of its store is handed over to housekeeping, and the management needs become so large that the computer silts up, not unlike large government and multinational organizations. Micro-electronics make possible a lot more distributive processing (small is beautiful), and devolution of management to secure information and sieve it before it gets to the centre.

Micro-electronics provide the capacity to collect, monitor and act on real-time data, but the problem lies in acceptance and adoption. How can these techniques be more easily and rapidly integrated into adequate instrumentation plant and other equipment? The techniques have been adopted quite quickly in traffic engineering but they apply equally well to all types of equipment, quality-control processes and system management, and there are many applications which have not been thought up because many engineers are unacquainted with the possibilities. It is therefore necessary to bring together the many facets of electronic development in discussions between industry users, public sector clients, researchers and so on to regenerate the imagination and entrepreneurial impetus that has been lost.

This is a cause of great concern and a key element of government policy. The Finniston Report (ref.1) shows that the rate at which patents are applied for has gone down more in the UK than elsewhere. Regeneration is important because only about 8% of the UK GNP is created in an indigenous way through coal, oil, tourism and the like and 92% is created by adding value to something which already exists. The stock in trade is therefore ingenuity in application - and it is under-exploited.

Professional engineers must give serious consideration to how best to devise a comprehensive specification of what is required. Developments are taking place in an unco-ordinated random manner.

The most demanding need is to make good and convenient products which are reliable, relatively maintenance-free and easily adaptable. Much thought has been given to the question of balancing reliability against the costs of maintenance. There is some assumption of certainty in determining benefits and costs, but there is nothing so certain as uncertainty and the DoT has launched considerable research work to produce evaluation methods which take explicit account of uncertainty.

The scientific definition of reliability is (1 - p) where p is the probability of failure, but the test to establish the minimum risk of failure which can be economically sustained is to establish whether or not further investment will lower the product of

risk of failure x economic consequences of failure

aggregated and discounted over a sufficient period of time by more than the value of the resources used.

This provides an approach to design which could be called economic limit state design. The aim of using control systems is to lower economically the risk of failure and the economic consequences by heralding the possible onset of failure and allowing preventive intervention. The method is aimed at minimizing the cost of being wrong and the cost of not being innovative is to lose markets.

Innovation may not come from the publication of papers but from those who are stimulated by discussion to think of a new idea.

One aim of the innovator is to judge whether or not he can achieve an end more cheaply. He must stay alive to his colleagues' problems and persist until solutions are applied.

REFERENCE

1. Finniston, Sir M. 'Engineering our future: Report of the Committee of Inquiry into the engineering profession.' HMSO, London, 1980, CMND 7794.

Date Due